Keeping

A Guide to Domestic
and Commercial Management

Katie Thear

Broad Leys Publishing

Keeping Quail: A Guide to Domestic and Commercial Management

First edition: 1987
Second edition: 1994
Third edition: 1998

Printed by Design and Print Ltd
Unit 5, Hamlet Industrial Centre,
Whitepost Lane, London E9 5EN.
Tel: 0181 986 8071. Fax: 0181 986 6541

Published by Broad Leys Publishing Company

A catalogue record for this book is available from the British Library

ISBN 0 906137 26 8

Outside front cover photograph: Fawn Coturnix quail (female)
Outside back cover photographs: Chinese Painted or Button quail (male)
Eggs of Commercial Japanese Coturnix

For details of other publications please contact the publishers:
Broad Leys Publishing Company,
Buriton House, Station Road,
Newport, Saffron Walden,
Essex CB11 3PL, UK.
Tel: (UK) 01799 540922
Fax: (UK) 01799 541367
E-mail: cgs@broadleys.com
Web site: http://www3.mistral.co.uk/cgs

Contents

Coturnix japonica, male, in an outside breeding pen

Meadow View Quail

QUALITY PRODUCTS WITH A TASTE OF THE COUNTRYSIDE

Church Lane
Whixall
Whitchurch
Shropshire SY13 2NA

Tel: (01948) 880 300

At meadow View Quail we have selected our quail not only for the number of eggs produced per bird, but also for the size of the egg. This has increased considerably over the last few years. Selection for meat production has run concurrently with this, also and possibly the most important selection for temperament. A quiet quail is a lot easier to handle, but also a lot more productive. Recently we have produced a great variety of colours and now have a large selection available.

Quail can be kept almost anywhere as long as it is dry and drought proof and will keep out vermin. These little birds can tolerate cold but NOT cold and damp conditions. A garden shed or a garage is ideal for a larger number, but just a few can be kept in a hutch on the lawn, especially in the summer months. We have designed and made a pen just for this purpose. Quail have a life expectancy of about 3 ½ years and can lay up to 300 eggs per year, but they will need 14 hrs of light per day to do this.

Hen layers mash or crushed layers pellets are ideal food, this and fresh clean water are all they really need. You can also feed them titbits of fresh greens. They are also very happy as scavengers in the bottom of the aviary, where they will tidy up slit seeds and lay eggs all summer. Quail are normally hardy and healthy, and thrive under many types of housing and husbandry (including a large fish tank in a town flat. Fish and water removed of course.) Each Quail will eat approximately 25 grams of food a day, at a cost of about 7.5p per week. They lay 5-6 eggs per week under ideal conditions, so quail are very economic producers of eggs. Each egg contains only 16 calories and are higher in protein, minerals and vitamins than a hens egg, but lower in fat content. Quail hens do become 'broody' and will hatch their own chicks, to do this they need a quiet are in which to make a nest of their own. Chicks take 17 days to hatch and will need chick crumbs for the first few weeks of life.

A Quail purchased from us comes complete with carrying box and enough food to enable them to settle in their new environment with a little stress as possible. If you should have any problems at all please do not hesitate to ring from the number above, (evenings best before 9:30).

Meadow View Quail

QUALITY PRODUCTS WITH A TASTE OF THE COUNTRYSIDE

Church Lane
Whixall
Whitchurch
Shropshire SY13 2NA

Tel: (01948) 880 300

Price List
January 1998

Our quail are selected for meat and egg production and also for temperament. Quiet quail are easier to handle and are also more productive.

Selected breeding trios   £ 9.00. }   carriage at cost
Pointer of lay pullets   £ 3.00. }   £18.50   min.
Fertile eggs £2.00 per dozen + carriage
5 tier cages (100 birds) all inclusive £252.65.
4 tier cages (80) all inclusive £216.50.

Special breeder ration produced to our formulation and includes, no antibiotics, hormones, growth promoters or artificial yolk colourants. This ration is a very high specification diet designed to produce high fertility, hatchability and chick viability, and is available as,

30p per 1b up to 10lb more than 10lb 25p per lb.

Runs suitable for Quail, Hens, Rabbits, Guinea Pigs, Bantams and even Ferrets £29.50.

Smoked and Pickled Quail Eggs £3.00 per jar.
Pickled Quail Eggs   £2.50 per jar.

Carriage on above,

1 or 2 jars   £2.50   per order
Up to 4 jars   £2.75   ``
Up to 6 jars   £3.00   ``
Up to 8 jars   £3.25   ``
Up to 10 jars £3.50   ``

Smoked salt (presentation jar 45gms)   £1.50 plus 60p p&p.
Smoked salt (sachet 28gms) £1.00 } 1 sachet P&P 26p
Chilli salt   (sachet 28gms)   50p } 2 sachet P&P 40p
Herby salt   (sachet 28gms)   50p } 3 sachet P&P 50p
                                    } 4 sachet P&P 60p
                                    }   Max. p&p   60p

Hawk food available at £5.00 per 10 (frozen) collected.
Contract smoking is also undertaken.
Prices include V.A.T where applicable.
Cheques made payable to E.F & F.E Reeves.

Introduction

"Are they birds of prey?"
"No, they're partridges."

(Two ladies looking at Coturnix quail at the Ardingly Smallholder Show, Sussex, 1986)

Quail are classified as game birds, sharing the same family as pheasants and partridges. Although there are over 40 different species of quail in the world, the number of species kept domestically or commercially, or as ornamental aviary birds, is quite small.

The book concentrates on the breeding and management of those species, notably the Coturnix laying strains, the Bobwhite, and the ornamental aviary breeds such as the Chinese Painted or Button quail. The book does not neglect the wider scale in relation to overall breed development and diversification, but it is primarily a practical book based on my own experiences of keeping quail.

The first edition of this book was published in 1987. It has sold all over the world and, as a result, I have received letters from many different countries. I am grateful to all those who have sent in suggestions and contributions, including the readers of the monthly magazine *Country Garden & Smallholding*.

This edition is as up to date as I have been able to make it. The reference section has also been updated so that anyone looking for sources of stock and equipment will be able to find the appropriate contacts.

I hope that quail breeders, producers, hobbyists, schools and agricultural colleges, and indeed anyone interested in these fascinating little birds will find something of interest within these pages.

Katie Thear, Newport, 1998

Fawn Coturnix, female. (See also the outside front cover photograph).

History

*"There went forth a wind from the Lord
and brought quails from the sea."*

(Exodus)

The earliest known representation of the quail is in Egyptian hieroglyphics, when the little bird was presumably common enough and important enough to merit a place in the alphabet. Carvings of the period indicate that it was caught in nets, along with other species, and represented an important part of the diet. In China, too it was represented in carvings and ceramics.

There are two closely related species in the wild, the Common or European quail, *Coturnix coturnix*, and the Eurasian or Pharoah quail, *Coturnix communis*. The chances are that they are derived from a common source. The Japanese quail, *Coturnix coturnix japonica*, has been bred from the wild form and is heavier. Its pedigree dates back to the 12th century when it was valued as a songbird in Japan! The Japanese is the type from which most farmed Coturnix are derived.

Stone Age man was apparently partial to quail meat! Remains of Coturnix quail have been found in early Upper Pleistocene rocks in Britain, notably in Chudleigh Cave and Kent's Cave in Devon, Hoe Grange Cave in Derbyshire and Kirkdale Cave in Yorkshire.

They make their appearance, too, in the literature, art and music of different periods. In the Bible, there is reference to how food was brought to the starving Israelites in the wilderness: *"There went forth a wind from the Lord and brought quails from the sea."* (Exodus). The Israelites feeding on quail are depicted in an 11th century Spanish manuscript (Vatican Library), as well as in the *Rohan Book of Hours*. The *Sherborne Missal* of 1400 has a good representation of a quail, while the *Pilkington Charter* (Fitzwilliam Museum, Cambridge) depicts a somewhat obscure one which could arguably be a partridge. The late 14th century English poem, *Pearl*, describes the quail's propensity for keeping still in face of danger: *"I stod as stylle as a dased quayle."*

Coturnix quail were kept as aviary birds by the early Greeks whose interests extended to their fighting as well as their culinary qualities. The island of Ortygia (the ancient name for Delos) is derived from the Greek name *ortyx* for quail. Dionysius and Aristotle refer to them, as well as the Roman writer Pliny who said their flocks were hazardous to shipping! Varro thought highly of them as a source of profit: *"Other birds fetch a good price, such as ortolans and quails."*

Shakespeare refers to the ancient practice of quail fighting in *Antony and Cleopatra*, when Antony bemoans the fact that Caesar's chances of winning are greater than his: *"His cocks do win the battle still of mine when it is all to naught, and his quails ever beat mine, inhooped, at odds."*

Early representations of the quail

Painted relief from the tomb of
Seti 1 at Thebes

Relief from the Temple
of Seti 1 at Abydos

Section of the Hunefor
papyrus scroll

𝓌

The quail in the
hieroglyphic
alphabet - 'ut'

Right: From the Transnotation
of Bird Voices by Athanasias
Kircher, 1650

Woodcut from Mrs Beeton's
*Book of Household
Management*, 1861

Woodcut from Thomas Bewick's
History of Birds, 1820

Shakespeare also uses the term quail as a reference to a prostitute, while in the USA, a 'San Quentin quail' is possibly of the same ilk. In Egypt the term quail is used to describe someone well fed. It was also commonly used as a dimunitive and familiar term of affection, as Donna Woolfolk Cross demonstrates in her novel of the 9th century, *Pope Joan*: *"Trust me, little quail."*

Beethoven utilizes the call of the quail in his music. See if you can spot it in the *Pastoral Symphony*! Leopald Mozart (father of Amadeus) also includes the quail in the *Toy Symphony*, while Althanasias Kircher produced a written notation in 1650!

The seventeenth century East Anglian poet, John Clare, could recognise the quail on a summer's evening: *"While in the juicy corn the hidden quail cries 'wet my foot' and hid as thoughts unborn."* As a true countryman, he did not confuse it with the similar looking but larger corncrake, which, *"utters 'craik, craik', like voices underground."*

Gilbert White, eighteenth century vicar of Selborne and author of that great classic, *The Natural History of Selborne*, makes several references to quail. He records hearing their calls between June 22nd and July 8th, 1774. In 1786, the calls were apparently emanating from *"the field next to the garden"*, while in 1787, the first call of the year was on May 20th *"at Rolle"*.

In 1820, Thomas Bewick illustrated the quail in his *History of Birds*, while, in 1859, Mrs. Beeton wrote: *"Quails are almost universally diffused over Europe, Asia and Africa, in the autumn, and returning again in the spring, frequently alighting in their passage on many of the islands of the Archipelago, which, with their vast numbers, they almost completely cover. It appears highly probable that the quails which supplied the Israelites with food during their journey through the wilderness, were sent by a wind from the southwest, sweeping over Egypt towards the shores of the Red Sea. In England they are not very numerous, although they breed in it; and many of them are said to remain throughout the year, changing their quarters from the interior parts of the country for the seacoast."* (It is unclear whether she is referring to the British Isles, or only to England - the latter being a frequent misnomer for Wales!)

Sir Herbert Maxwell wrote: *"There was a moderate immigration in 1893"*, commenting that, *"the fecundity of the species must be prodigious: millions being taken in nets during their spring and autumn migrations across the Mediterranean."* These went, *"to the great cities of Europe, where hundreds of thousands are consumed at the tables of wealthy persons."*

The effects of intensive farming have been to reduce the numbers of wild Coturnix quail which arrive in Britain. At one time, they were widespread as summer visitors. Now, they are rare immigrants arriving between May and July from Africa, via the Mediterranean and Northern Europe. In 'good' years they are to be found in cereal and hay fields in Hertfordshire, the North Wessex Downs, the Oxfordshire and Berkshire border, and Wiltshire.

The quail's retiring habit has given rise to the verb *to quail*. The Reverend Patrick Bronte makes use of it in a letter to Ellen Hussey in 1885, when he refers to Mrs Gaskell's undertaking to write the life of his daughter Charlotte: *"No quailing, Mrs Gaskell! No drawing back!"*

In recent years, a considerable Coturnix quail industry has grown in the United States, Japan, Italy, France and Britain. This is to cater for the gourmet interest in quails' eggs and delicatessen table birds. In Japan and other countries, it also has the dubious distinction of being bred for laboratory testing.

Worldwide, there are thirteen sub-groupings of quail. The Scaled or Blue quail, for example, is indigenous to the desert areas of southern USA and northern Mexico, although it is available as an aviary bird in Britain. So, too, is the California quail, *Lophortyx californica*, the state bird of California where it is a relatively common sight. I saw several of these atttractive little birds, on the edge of Lake Cachuma. The Brown quail, *Coturnix australis*, is found in the wetland areas around Brisbane. Compared with the total number of quail breeds in the world, only a small number are kept and bred domestically. *The Quail Group* of the *World Pheasant Association* monitors the situation and there are also clubs catering for specific breeds such as the Chinese Painted quail.

The Bobwhite quail is kept as a farmed table bird in some areas of the USA, and also as a 'managed habitat bird' for the hunting sector, in a similar way to that of pheasant rearing in the United Kingdom.

The Chinese Painted or Button quail, *Coturnix chinensis (Excalfactoria chinensis)*, is widely kept as an aviary bird. Traditionally it has been kept as a ground clearer when flying birds such as finches and doves spill seeds on the aviary floor. In recent years, with the growing popularity of butterfly houses, it has also developed a role as a spider catcher, to prevent butterflies being caught in the webs. More recently still, it has become popular in its own right, as a small and beautiful aviary bird, or pet.

There have been many references to quails, but perhaps the last word should go to a schoolboy who sent me the following poem about the quail, and which I was pleased to include in the first edition. I don't think that any of the writers of ancient civilizations have bettered his description:

> *"Short and plump,*
> *Timid but inquisitive;*
> *Dull but pretty,*
> *Not very clever,*
> *But a long way from silly."* (Alex Smith)

About the quail

"We loathe our manna and we long for quails."

(Dryden, 1682)

There is a wide variety of breeds, adapted to varying climates and conditions throughout the world, but in this brief look at the general characteristics of quail, I have concentrated on the more common breeds.

Most quails are birds of the undergrowth which, depending on the area, may be tall grasses, dense bush, shrub thickets, overgrown fields, meadows, plantations or savannahs. They are essentially shy, retiring birds which will 'quail' with fear into the shadow and security of the undergrowth. They stand motionless. When really disturbed however, they will break cover like pheasants, and fly straight up in the air with a characteristic whirring of wings.

Characteristics of the quail

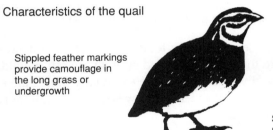

Long, sharp beak ideal for picking up grains and insects

Stippled feather markings provide camouflage in the long grass or undergrowth

Sharp eyesight provides protection against predators such as hawks and crows in the wild

Strong claws scratch the soil to reveal insectivorous food

Ground orientation

Most are ground-orientated in that they spend most of their time on the ground in the wild, except when they are migrating or disturbed in some way. Some are more ground orientated than others. The Chinese Painted or Button quail, for example, is like a small mouse scuttling about on the ground, resorting to occasional short flights where necessary. The Bobwhite, by comparison, likes to fly and perch, and does not scratch as much as Coturnix. It is important, therefore, to make provision for such behaviour within an aviary system. Clumps of grass or shrubs are needed to provide cover, with a couple of branches for perching.

The stippling effect on the plumage provides good camouflage. The lines blend in with those of tall grasses, so clumps of vegetation in the aviary are well received. I once dropped a Japanese quail outside when I was transferring him into an outside grazing pen. He immediately disappeared into some tall grasses

The Bobwhite quail is a perching bird and needs adequate flying and perching space. A hard surface such as this (shown for this photograph only) is unsuitable.

near a copse of trees. Despite a search, I never found him, yet he could have been just a few feet away. I like to think that he joined up with migratory wild quail which are periodic visitors to my area, and that he eventually joined the migration back to Europe and Africa. The chances are, however, that he fell foul of the many pheasant shooters in the area.

Toes and claws

Most breeds of quail have long toes and claws, an adaptation for scratching for insectivorous foods. In an aviary there is usually provision for this, but if there is only a concrete run, do make sure they have an area where fine soil, sand or

Claw clipped to avoid cutting blood vessel

wood shavings can be made available. The Bobwhite's strong claws are particularly adapted for perching, as may be seen in the photograph above, and a hard flat surface is inappropriate, although acceptable for Chinese Painted quail. Scratching conditions are vital for Coturnix breeds, not only to cater for their instinctive need to scratch, but also to keep the claws in trim. In housed conditions it may be necessary to cut them. Hold the bird gently in one hand, using the fingers to keep the limb steady, then clip the claws with nail clippers. Do this in a good light so that you can see where the blood vessel within the claw ends, so that the cut is made beyond it. It is painless, and nail clippers can be used.

Stippled feather markings (lines) seen on this Japanese quail, provide camouflage and protection. Once in the long grass the bird stays still and is difficult to see.

The beak

The quail's beak is long, pointed and sharp, ideally suited for pecking small insects and grains, or for shredding small pieces of vegetation. It can also be an aggressive weapon if one quail decides to attack another. The males will fight to the extent that they cause considerable damage.

In some intensive units, quails are beak trimmed; the upper beak is trimmed back so that it is shorter than the lower one, but I find such a practice repugnant.

Another practice, more common in the USA than in Britain, is to use beak rings. These are open rings which have the ends inserted into the nostrils and then pinched together with pliers. Again, I would not advise such an approach.

In the wild, the beak is kept relatively short by access to a range of natural materials. In captivity it is a good idea to provide dried cuttlefish for this purpose, as is commonly done for canaries and budgerigars. Most pet shops sell it. Make sure that it is positioned firmly so that there is adequate friction when the beak is rubbed against it.

Calls

The question is often asked - how noisy are quail? As with all things, only a relative answer is possible. As a generalization, males are noisier than females, which confine themselves to subdued little chirrups and soft 'tic-tics' or cooings. The male Coturnix has a rasping chirrupping crow, rather like the harsh cry of the magpie, and is altogether louder and more penetrating. Traditionally, this has been described as sounding like 'wet-my-lips' or 'wet-my-feet'. Athanasias Kircher, in 1650, decided that it was saying 'bik-e-bik', and produced a musical notation to illustrate it (See page 8).

The photograph opposite shows a Tuxedo Coturnix male adopting the characteristic upright stance and open-beaked expression when letting the world know of his whereabouts. This uptight stance is also the one that was depicted in the Egyptian hieroglyphics.

The Bobwhite male is also loud, but I do not find his call as aggravating as that of the Coturnix. It is basically a three-note, full-bodied piping, whistling sound. The female has a quieter version of the same call, with more of a tendency to chirrup.

The Chinese Painted male has a melodious, rather wistful piping whistle. It is almost minor-key, with an evocative atmosphere of the jungle in it. The female is quiet, apart from the occasional and busy 'chic-chic' as she darts about looking for insects.

If there is likely to be a problem of near neighbours complaining about 'noise nuisance' (which is a recognised offence in the UK), it may be better to keep Chinese Painteds, or if Coturnix quails are really wanted, to do without the males, although this can be difficult to organise.

Head banging

In the wild, Cotumix quail will break for cover when disturbed, and fly straight upwards. This is the same pattern of behaviour as that seen in pheasants.

In housed conditions the same pattern is also found. The birds have a tendency to fly straight up and dash their heads on the roof, sometimes causing injury. It is apparent all through the year, but is much more

Soft netting protects the birds from banging their heads on the roof

so in the period leading up to the breeding season, when the migratory urge is at its peak.

The only way to deal with this is to stretch soft netting just below the solid roof so that they do not hurt themselves.

Tuxedo variety of the Coturnix quail (male). Note his upright stance when calling.

Preening

Like all birds, quails will preen themselves regularly, often standing in a patch of sunlight to do so. This ritual of cleaning the feathers with the beak is also a sign that the bird is healthy. A quail which does not preen itself, but has a tendency to stand, looking morose, should be checked, because it is a sure sign that all is not well.

American Range variety of Cotrurnix quail on the author's conservatory table. Note the large feet in comparison with its size, an adaptation for scratching for insectivorous food.

Taming

As a general rule, males should be kept separate to avoid fighting and inter-breeding, but on a small scale, where there is less pressure of numbers and general stress, they may live together quite happily without fighting. On a small scale, birds do become tame.

I frequently allowed up to seven different males to share the same aviary and run in the summer, with occasional browsing periods on grass in a movable run. In winter, they were housed in canary breeding cages in the conservatory, but were frequently released to enjoy the freedom of the conservatory.

I did not let the breeds interbreed, of course. When I wanted to collect eggs for incubation purposes, I confined the male and females of a particular breed together and kept them separate from the others until I could be sure that the eggs were pure.

English White variety of Coturnix quail

I must emphasize that my birds were probably demonstrating behaviour different from the norm, because they were so tame. On sunny winter days, I often worked at a table in the conservatory. One or more of the quail would fly up and see what I was doing, often settling down and going to sleep a few inches away. As a matter of interest, there is an American book called *That Quail Robert* by Margaret Stranger which is a fascinating account of how the author tamed a Bobwhite quail after hatching him.

Tuxedo variety of the Coturnix quail. Ideally, his white waiscoat should be clearly differentiated from the dark feathers, rather than being 'patchy'.

Breeds

"The quail whistle about us their spontaneous cries."

(Wallace Stevens, 1923)

If we 'place' the quail in its relative position in the bird world, it is found in the order of *Galliformes*. This is a group which includes game birds and domestic fowl, but not waterfowl. Narrowing the classification further, it is a member of the *Phasiandae* family, a category it shares with pheasants and partridges.

Class: AVES (Birds)
|
Order: GALLIFORMES (Game birds and Fowl)

Family: TETRAONIDAE   Family: PHASIANIDAE

Grouse Ptarmigan Capercaillie   Pheasants Partridges Quail

Within the quail group, there are thirteen sub-groupings as indicated:

QUAIL
- Anurophasis species
- Callipepla species
- Colinus species
- Coturnix species
- Cyrtonyx species
- Dactylortyx species
- Dendrortyx species
- Odontophorus species
- Ophrysia species
- Oreortyx species
- Perdicula species
- Philortyx species
- Rhynchortyx

Each of these sub-groupings has one or more representative species and some also have sub-species, demonstrating the genetic diversification in different geographical locations. There are over forty species of quail in the world, with more than twenty currently being bred in captivity. The *Quail Group* of *The World Pheasant Association* has an ongoing survey of what breeds are being bred, providing valuable recordings for conservation work.

Sub-groupings of quail

Anurophasis
Anurophasis monorthoryx — Snow Mountain quail

Callipepla
Calipepla californica (Lophortyx californica) — Californian quail
Calipepia douglasii — Elegant quail
Calipepia gambelii — Gambel's quail
Calipepla squamata — Scaled quail

Colinus
Colinus cristatus — Crested Bobwhite quail
Colinus nigrogularis — Blackthroated Bobwhite quail
Colinus virgianus — Bobwhite quail

Coturnix
Coturnix coromandelica — Rail quail
Coturnix coturnix — Common or European quail
Coturnix communis — Eurasian or Pharaoh quail
Coturnix japonica — Japanese quail
Coturnix delegorguei — Harlequin quail
Coturnix chinensis (Excalfactoria chinensis) — Chinese Painted or Button quail
Coturnix pectoralis — Grey quail
Coturnix novaezeelandiae — New Zealand quail
Coturnix ypsilophorus — Brown (Swamp) quail

Cyrtonyx
Cyrtonyx montizumae — Mearns quail
Cyrtonyx ocellatus — Ocellated quail

Dactylortyx
Dactylortyx thoracicus — Singing quail

Dendrortyx
Dendrortyx barbatus — Bearded Tree quail
Dendrortyx leucophys — Buffycrowned Tree quail
Dendrortyx macroura — Longtailed Tree quail

Odontophorus
Odontophorus atrifrons — Blackfronted Wood quail
Odontophorus balliviani — Stripefaced Wood quail
Odontophorus capueira — Spotwinged Wood quail
Odontophorus columbianus — Venezuelan Wood quail
Odontophorus dialucos — Tacarcuna Wood quail
Odontophorus erythrops — Rufousfronted Wood quail
Odontophorus speciosus — Rufousbreasted Wood quail
Odontophorus gujanensis — Marbled Wood quail
Odontophohus guttatus — Spotted Wood quail
Odontophorus hyperythrus — Chestnut Wood quail
Odontophorus leucolaemus — Whitethroated Wood quail (Blackbreasted Wood quail)
Odontophorus melanonotus — Darkbacked Wood quail
Odontophorus stellatus — Starred Wood quail
Odontophorus strophium — Gorgeted Wood quail

Ophrysia
Ophrysia superciliosa — Indian Mountain quail

Oreotyx
Oreotyx pictus — Mountain quail

Perdicula
Perdicula argoondah — Rock Bush quail
Perdicula asiatica — Jungle Bush quail
Perdicula erythrorhyncha — Painted Bush quail
Perdicula manipurensis — Manipur Bush quail

Philortyx
Philortyxyx fasciatus — Banded quail

Rhynchortyx
Rhynchortyx cinctus — Tawny faced quail

Some of the author's quail in a movable run. This is temporary accommodation, giving access to grass during summer days. They are tame and show no inclination to fight.

The table opposite shows the 13 sub-groupings of the quail family. Some, such as *Coturnix*, *Colinus* or *Oreortyx*, have one or more species. These, in turn, may have their own sub-species, such as different colour varieties.

Size

Quail vary in size, from the largest Longtailed Tree Quail, *Dendrortyx macroura*, at 36cm (14"), to the smallest Chinese Painted quail, *Coturnix chinensis* (*Excalfactoria chinensis*) at 12cm (4.5"). They are found in a wide range of climates, but all share similar characteristics of being shy, quick and with a tendency to hide in ground cover such as long grass or other vegetation.

Life span

There is little information available on this aspect, but Altman and Ditmer, in their report, *Growth including Reproduction and Morphological Development*, 1962 (USA), say that the record life span of a quail is ten years. Commercially, it is between 1-2 years.

Some breeds of quail showing the relative difference in size

Common European quail
Coturnix coturnix (wild species)

Coturnix laying quail
Coturnix coturnix japonica (developed strain)

Chinese Painted or Button quail
Coturnix chinensis (*Excalfactoria chinensis*)

Bobwhite quail
Colinus virginianus

Californian quail
Callipepla californica

Mountain quail
Oreortyx picta

Wood quail (Starred variety)
Odontophorus stellatus

Two of the author's Coturnix laying quail in 'natural' conditions to which they are adapted.

Coturnix quail

This is the most common type in captivity worldwide. It is essentially the same bird that the Ancient Egyptians knew and that Mrs. Beeton would have recognised as one of the *"feathered game which have from time immemorial given gratification to the palate of man"*. When reference is made to 'quail' in general, this is the bird in question. In Europe, it has been known as Common quail, European quail and Mediterranean quail. The old English names, 'wet-my-lips' or 'wet-my-feet', are reference to its familiar call.

The name Pharaoh's quail is linked to its Egyptian origins, while in the USA, early settlers referred to it as German quail, no doubt because German settlers brought it with them. Eurasian quail is a reminder of its considerable geographical distribution. Other names used in the past are Nile and Mediterranean quail.

Quails have been developed and to a certain extent domesticated, although the word 'domesticated' is used advisedly for they are still essentially 'wild' in

The distinctive eye stripe is common to both sexes of Coturnix quail varieties.

their form and behaviour, and are of course, designated as game birds. Although selective breeding has taken place to produce egg laying strains or birds more suited to the table, the development is nowhere near as marked as it is with domestic fowl.

No-one can know for certain how the various breeds and sub-species developed, but it is generally acknowledged that the Coturnix types are based on the Common European quail, *Coturnix coturnix*, Pharaoh or Eurasian quail, *Coturnix communis*, and Japanese quail, *Coturnix japonica*. The latter was first recognised in the nineteenth century as a separate breed in the wild, although it had been bred by the Japanese as a singing bird as early as the 12th century. In recent years, the Japanese have developed more productive commercial strains for the table, and for the laboratory, which are also called Japanese quail.

Selectively bred laying strains developed from *C. coturnix*, *C. communis* and *C. japonica* are often impossible to distinguish. For convenience, it is easier to refer to them merely as Coturnix laying quail. In America however, they tend to call Coturnix layers, Pharaoh quail. In Japan they understandably call them Japanese quail. In Britain where, as always we compromise, we call them all these things. It can be confusing!

The male Coturnix grows to a maximum of 16cm (6.5"), while the female is slightly larger at an average of 18cm (7.5"). Both sexes have dappled dark brown, buff and cream striated backs, paler underbellies, breast and flanks. In the female, the markings are less pronounced, while the male's chest is reddish brown. This particular feature enables sex identification to take place at around 3 weeks of age. Before then, it is difficult to do so.

In both sexes there is a distinctive light stripe above the eye, and a white

24

Left to right: Tuxedo Coturnix, Bobwhite and Fawn Coturnix.

collar, although this may be diminished in the female. The beak is yellow-brown to dark olive-brown, the legs pinkish yellow and the eyes dark brown.

The Coturnix laying quail has been selectively bred for commercial production to a limited extent, and the breeders have, in turn, given their names to particular strains. *Curfew* have developed their own *Crusader* strains, while in the USA, *Marsh Farms* produced *Marsh Pharoah* strains.

Coloured varieties of Coturnix

As well as the normal or commercial type of Coturnix laying quail, there are several varieties which show distinct colour variations and markings.

Manchurian quail: The alternative name for these is Manchurian Golden because of their colouring. They are essentially the same breed as Japanese or Pharoah, but have been developed as a separate variety with golden colouring. Markings are similar but the overall hue is lighter and more golden.

Range: The overall colouring is dark brown, so that some people refer to them as Brown quail. This is a mistake however, for the Brown quail is the name normally given to the Australian breed which is bigger and more greyish in appearance. However, there is a similarity and it is possible that Australian settlers introduced the Common quail to that Continent, with subsequent isolated development producing apparently different species.

The markings of the Range are essentially a lighter brown body colour overlaid with darker-brown, almost black pencilling, along with a certain amount of dark grey feathering on the back. The latter is a point of comparison with the Australian Brown quail.

In America, the Range Coturnix, is referred to as the British Range, while in the UK, we call it the American Range! The overall appearance is dark-brown while the striped head markings are similar to other Coturnix breeds. However, the white eyebrow stripe and white throat markings are virtually absent. Beak and legs are olive-brown and eyes, dark brown.

Fawn: This is the variety shown on the outside front cover and is one of my favourites. Essentially like all the other Coturnix breeds, the overall impression of the Fawn is a lovely warm pinkish-brown. The fawn feathers are pencilled with white and the white eyebrow lines are present, although not as strongly defined as in other breeds. Beak and legs are light pinkish-brown, and eyes are dark brown.

There is no colour difference between the sexes, although as in other breeds, the female is slightly bigger than the male.

English White: Good specimens of these are completely white, with no discernible markings, other than the merest hint of eyebrow lines on the head. It is common, however, to have odd patches of black, and a look at the photograph on page 17 indicates that mine do have slight markings on the back of the head. Breeders who are aiming for perfect, all-white specimens can breed this out with careful selective breeding. Beak and legs are pinkish brown and eyes are dark brown.

Male and female are identical, although the female grows to a larger size, particularly noticeable once breeding starts.

Tuxedo: This is an apt name for a bird with a smart white waistcoat to contrast with its dark brown overcoat. The colour of the back feathers is identical with those of the Range, indicating the close connection with that variety as well as with the English White. The ideal markings are clear white face, chest and belly, with brown back, tail and crown. In good specimens, the brown and white feathering is neatly demarcated, but it is common to find patches of white where the brown should be, and vice versa. If you look at the photograph on page 18 you will see that one of my Tuxedos has the brown and white plumage interspersed.

Commercial producers of eggs and table quail will concentrate on the Coturnix laying strains bred commercially. On a smaller scale, there is no reason why the interested reader should not go in for the colour variations of Coturnix. They are often prettier and are usually good egg producers. Egg production levels can, of course, be increased by selective breeding, and it is really the particular 'strain' which is relevant in this respect.

The Range (also called American Range or British Range) Coturnix has much darker plumage than the normal Coturnix laying quail.

In addition to the normal Japanese Coturnix laying quail, I have kept Range, English White, Fawn and Tuxedo. The first do lay more than the others but the difference is not that marked and would only be unacceptable in a commercial egg production unit.

On a small scale, many breeders find that concentrating on coloured varieties is not only more interesting, but they are able to sell breeding pairs or trios of stock in addition to, or instead of eggs.

The author's Bobwhite quail (male) above and right

Bobwhite quail

The Bobwhite quail, *Colinus virginianus*, is a breed which has its distribution mainly in North and Central America, although its indigenous habitat is east of the Rockies in mid-west and southern USA. Alternative names are American quail, Northern Bobwhite and Partridge quail. The latter name is an appropriate one for they are more like partridges than quail. They are not migratory birds, like the Coturnix, preferring to stay in the same locality.

In the USA it is primarily regarded as a 'managed habitat game bird' for the benefit of hunters. Over-hunting, change of land use and the effects of intensive farming have all had a negative effect on population numbers. The *Wildlife & Fisheries* department of *Mississippi State University* estimates the loss to be 70% in the last thirty years. Where it is raised for meat, the Eastern variety of Bobwhite is heavier and therefore more appropriate.

In Britain, the Bobwhite quail is regarded as an aviary bird. There are over twenty sub-species of the breed, with some, such as the Masked Bobwhite, being so rare as to be on the endangered species list.

The Bobwhite is a bigger bird altogether than the Coturnix layer, with the male reaching 25cm (9.5") and female 27cm (10.5"). By comparison with other game birds, however, it is relatively small.

The back, tail and crown of both sexes is dark brown, while the chest, belly and flanks are lighter, with black and white striations. A white stripe covers the eyebrows and, in the male, there is a white patch under the chin. In some females, this patch is absent or reduced, being replaced by buff markings. The overall colour effect is less bright in the female.

The beak is greyish-brown, legs yellowish-brown and eyes dark brown. The Bobwhite is a most attractive bird. I kept a pair in an aviary and they proved to be easy to tame as I have already mentioned. They have a greater tendency to fly and perch than other breeds of quail, so they really need perching facilities. They can sometimes prove to be aggressive with smaller breeds, but I never found this to be so with mine which cohabited quite happily with Chinese Painted quail and some of the coloured varieties of Coturnix. I should emphasize however, that I kept relatively small numbers of each breed, and succeeded in taming my breeding stock to the extent that they would come when called. Anyone keeping larger numbers, or on a commercial basis, would obviously keep them separate, and would have less time for conversations.

Chinese Painted quail

The Chinese Painted quail, *Coturnix chinensis*, is probably the most widely kept of the ornamental aviary breeds. It is easily the prettiest and most colourful, and in recent years has been much utilized as a spider catcher in butterfly houses. Anyone who has ever visited one of the increasing number of butterfly breeding establishments, such as that at Syon Park in Middlesex, will have seen these busy little birds scuttling through the ground vegetation of the greenhouses, beady eyes on the lookout for unwary spiders and ground insects.

It goes by several names, depending on the part of the world in which you happen to be. In Europe we refer to it as the Chinese Painted quail, while the Americans incorrectly call it the Button quail. (The latter is not a quail at all, but a bird related to the rail, of the family *Turnicae* in the *Gruiformes* order). It is designated as *Excalfactoria chinensis* by some, while the Linanean classification *Coturnix chinensis* is used by others.

There are several sub-species. Paul Johngard lists a total of ten:

Coturnix coturnix chinensis	*C. c. lineata*
C. c. palmeri	*C. c. lineatula*
C. c. australis	*C. c. colletti*
C. c. trinkutensis	*C. c. lepida*
C. c. papuensis	*C. c. novaeguinea*

Found in China, India, Sri-Lanka, South Africa and Australia, the Chinese Painted quail is also known as the Blue Breasted quail. The sub-species are similar, but with slight variations. The Australian or Eastern King quail, *C. c. lineata*, for example, has more distinct lines on the plumage, hence the *lineata* designation. The African Painted quail, *C. adonsonii*, is also closely related.

No-one knows precisely what are the origins of many of the birds kept and bred in captivity. It is likely that what we know as the Chinese Painted quail is any one of, or a mixture of, several sub-species. Only extensive DNA sampling could indicate the precise relationship. In recent years there has also been considerable breeding for colour variation.

The Chinese Painted quail is the smallest of the quail breeds, with the male reaching 12cm (4.5") and the female 13cm (5"). In appearance it is compact and round, with a mouse-like way of scuttling about. The male is far more colourful than the female, with a brown and blue flecked back and crown. The breast and tail feathers are reddish brown, while the chin and throat have distinct black and white striping, like a smart crescent collar. The female is less flambouyant, with an overall mottled brownish hue from the fine black and white specks. Her back is slightly darker than her abdomen. She has a white patch on the throat but no barring. In both sexes the beak is black and the eyes are brown.

Chinese Painted (Button) quail, the smallest breed and a popular choice for the aviary.

Colour varieties of Chinese Painted quail

The different varieties are called by the colour itself, or referred to as 'coloureds' or 'mutations', eg, Blue, Blue Coloured, or Blue Mutation.

Silver: This is the most common mutation and has all the feathers in varying shades of light pastel grey.

White: These are all-white, although not albinos. There may be a few coloured feathers in some birds, but the aim is to produce a snow white effect.

Red-breasted: The face is almost black with a fine white line around the eye. The red area extends from the vent, across the breast to the throat.

Red Breasted Silver: Here the pastel grey plumage contrasts beautifully with the pinkish red breast.

Fawn or Cinnamon: Developed in Australia, this is called Fawn in Europe and Cinnamon in the USA, although differences are now being developed, eg, the Fawn is slightly darker. There is also a Blue-Faced and a Red-Breasted Fawn, as well as a Smoky version.

Blue: The plumage is a dark, overall blue.

Black: This is an even darker blue, approaching black.

Ivory: Lighter than Silver, it has an overall ivory hue, and the male's breast is grey.

Golden Pearl: Originating in Europe, this has yellow feathering with light brown barring. There is also a Fawn or Cinnamon Pearl, and a Blue-Faced Pearl. No doubt there will continue to be other colour variations produced.

31

Commercial quail in battery cages. Photo by courtesy of *Poultry World*.

Housing

"Floor pens provide the birds with a greater freedom of movement, and the opportunity for wider social interaction. Less stereotypic behaviour is seen in floor pens than in cages".
(Code of Practice for Housing & Care of Animals in Designated Breeding Establishments. 10: Quail - Home Office)

Quails are fairly small birds and their needs are modest, but they still need proper housing. Whatever form it takes, it must provide shelter from the elements and predators, adequate warmth and ventilation, and a clean, hygienic environment. It's worth taking a look at the optimum requirements for adult quail. This does not mean that they are crucial, but they can provide useful guidelines:

Temperature: 16-23°C
Relative humidity: 30-80%

The system of housing depends very much on the type and scale of the quail enterprise, as well as on the type of quail being kept. A commercial system for Coturnix layers or table birds will obviously be more intensive than that where birds are kept primarily for interest. A small-scale enterprise may utilise small houses and runs. The more exotic breeds will usually be in aviaries, either indoor or outdoor, with attached house. Bobwhite quail, for example, must have aviary conditions with perching facilities.Whichever system is used, the salient point is that quails are not winter hardy and will need adequate housing.

Most commercial enterprises will use either a cage system or a floor-rearing one, or perhaps a combination of both. The former is where purpose-made cages, equipped with automatic watering facilities are used. The latter is where a building with a concrete floor is furnished with wood shavings, feeders and drinkers, and the birds can roam at will within the confines of the house. A concrete floor is essential, and the building needs to be substantial enough, not only to deter rodents but also to provide draught-free and well-ventilated, sheltered accommodation.Ventilation by roof ridge and side window is ideal, as a through current of fresh air is provided. The arrangement of this will vary depending on the type of house. Very large houses will use fans.

Lighting in a house is not only necessary for efficient managment, but it will also provide a stimulus for winter eggs, although table birds should not be given extra light. Heating in winter is not necessary as long as there is enough insulation. Newly-hatched quail will naturally have localised heat in brooders until they are hardy. As there is so much variation in the types and scale of housing, it is a good idea to take a more detailed look at the options.

Cage systems

Cages are used by most commercial units where eggs are being produced. Some people, myself included, dislike cages on humanitarian grounds. There is no doubt that they make life easier for the producer, and egg collection is facilitated, but the birds' movements are restricted and foot problems are common. The use of wire bottoms for the cages facilitates cleaning and prevents a possible build-up of disease, but does mean that the birds are unable to scratch in the way they do in unrestricted conditions. Where cages are used, it is recommended that they are coated in plastic to reduce problems of this nature.

There are also 'deep-litter' type cages available in addition to the normal type, and these are favoured by many people because they provide a more natural flooring, enabling the birds to scratch.

Quail cages can be stacked on top of each other, to an overall height of 1.9m (6'3"). Each unit has a drop-tray for ease of cleaning, and feeders and drinkers are provided at the ends and sides of the cages. Automatic watering in the form of header tank, pipes and clips can provide a round-the-clock supply of fresh water. These particular cages can be used in association with *Curfew* tiered brooders. Quail chicks can then be raised from day-olds in the brooders until they are three weeks old when the brooders are switched off. The brooder area and run can then be used for the main runs.

The cages are 50 x 60 x 28cm (20 x 24 x 11"). Each cage comes fitted with a feeder, and manually filled or automatic drinkers are easily fitted. The floor is plastic covered mesh which is easy on the feet, and eggs as they are laid, roll forward for collection. The cages can be stacked four or five high and there is a removable droppings tray beneath each one. Each cage, in theory, can hold 15 quail, but it is not necessary or advisable to have as many as that. In summer, each cage can be lifted out quite easily be placed on grass in a sheltered area so that the quail have access to the grass which comes up through the mesh. One small producer I met does this regularly, placing her cages outside, on a different area each day so that it keeps down grass in her adjoining paddock. The cages have a windbreak in the form of a line of straw bales placed in the way of the prevailing wind. If the weather is particularly windy, cold or wet, cages are left inside. This practice, according to the owner, is a reasonable compromise, for she shares my dislike of batteries in general, but can see the value of them if used with a little imagination. The supplier of the cages confirmed that several producers use his cages in this way, adding that most quail producers are in fact, small scale, and prefer to utilise small scale practices.

It is also possible to make your own cages, and welded mesh is suitable. A protective strip for the edges of cut welded mesh is available by the metre.

A system of indoor rabbit hutches adapted by the author to provide Coturnix quail housing.

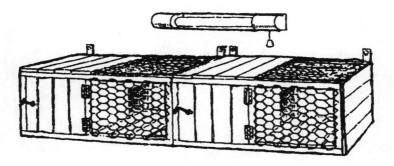

The covered areas provide sleeping accommodation while the wire mesh on the roof of the living area allows adequate light to enter to ensure winter egg production. This was winter housing only. In the summer months, the birds were in movable runs on grass.

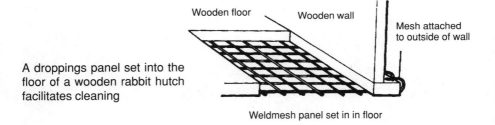

A droppings panel set into the floor of a wooden rabbit hutch facilitates cleaning

Wooden floor   Wooden wall

Mesh attached to outside of wall

Weldmesh panel set in in floor

Some people prefer to make their own cages using a wooden framework. Other alternatives are to adapt rabbit hutches or wooden bird breeding cages with solid wooden floors, sides and roofs, and a mesh cage front. These require a greater degree of management because the litter needs to be cleared out regularly, but it is not an enormous task on a relatively small scale.

Some of these wooden cages are often equipped with a sliding floor section so that cleaning is relatively simple. A section of weldmesh panel can also be inserted in one area of the floor to facilitate cleaning. I adapted a bank of rabbit hutches in this way and they worked well. The details are shown in the illustrations above.

It is important to bear in mind that, while female birds can be housed together without any problems, males will fight, sometimes to the death, and must be housed separately. On a large scale, this can be difficult to organise.

A floor system in protected housing. Photo by courtesy of *Poultry World*.

Floor systems

A floor system is set up inside a building where the door has an interior flight entrance. This is essentially to stop birds escaping when you open the door, and works on the 'air-lock' principle. In other words, you close the outside door before opening the interior one. It need not necessarily be a door; it could be strips of wood or wire mesh strategically placed to provide a barrier.

It must be emphasised, that Coturnix quails can be great escapers and, as referred to earlier, can be impossible to locate once they get outside and find some natural cover.

Concrete floors are also essential for such a system, for rats are highly intelligent and devious in their efforts to gain entrance into an area where there are vulnerable young birds. Wood shavings or sawdust provide the best litter material. They are clean, absorbent and also absorb smells, but do ensure that they have been produced for bird use, in case there are toxic residues from the timber.

Suspended feeders and drinkers are probably best in this sort of situation, for they are less likely to have litter scratched into them.

A less intensive floor system for those who like to give their birds more freedom, is one which has the inside run extending outside. An example is shown in the illustration on page 39. Here, the birds range on the concrete floor with wood shavings inside the house and also on a concrete run outside because the run is rat-proof, but droppings can be swept with a broom, and if necessary, hosed down when the birds are confined inside by closing the pop-hole. Naturally, woodshavings are not used in the outside run, but it may be a good idea to provide a shallow, dustbath area of fine sand so that they can take dustbaths in fine weather.

The outside run is enclosed by stout wire mesh of a gauge too fine to allow rodents to enter and the bottom section has protective boards which help to protect against the wind. It is about 90cm (3') high with a gate for human access, making run clearance easier. The provision of a certain amount of natural cover will be welcomed by the birds. Some conifer branches placed in one corner are suitable, or conifer trees in pots. Bracken cuttings and various logs have also been used to good effect. Even if the run is completely contained within a building, the provision of such cover is appreciated. The fact that birds are kept in artificial conditions does not mean that their natural instincts are diminished.

The great disadvantage of floor systems from the viewpoint of the commercial egg producer, is that quails will not use nest boxes, and will lay their eggs anywhere. This does mean that they are more prone to damage and dirt. Some egg producers raise the young on floor systems then transfer them to cages when they begin to lay.

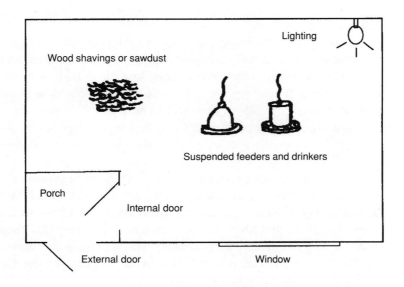

Here an 'open house' system is used, where all the birds are in the same area. The porch with its external and internal door prevents any escapees, and also provides draught protection. Artificial light to provide a maximum of 15-16 hours of natural/artifical light is necessary for layers. Table birds do not need extra light.

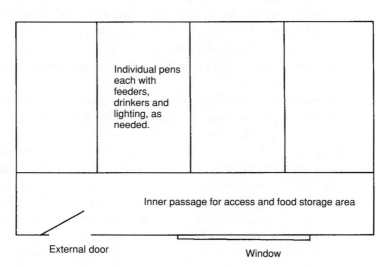

Here the area is divided into several pens with access from an internal passage which not only acts as a porch, but also doubles as a food storage area.

Two ways of using a floor system of rearing

A series of interlocking posts and game netting can be used to make internal pens in a building where a floor system is used. This simplifies management and prevents mass panic when the birds crowd in the same direction at the same time. It can also be used to make internal divisions in an aviary, as well as temporary outside pens in the summer. Another application is as an internal door in a floor system building, preventing birds escaping when the main door is opened. Photos by courtesy of *Agriframes Ltd.*

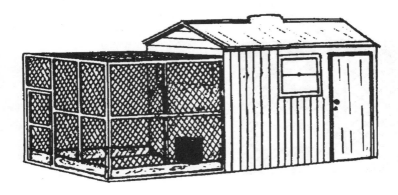

A less intensive floor system where quails have access to an outside run via a pop-hole.

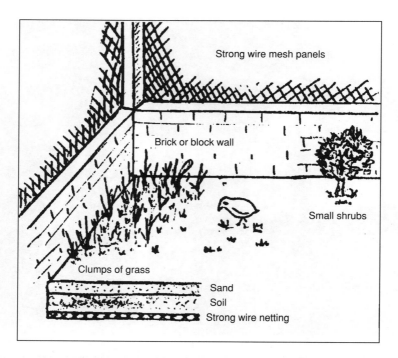

Strong wire mesh panels

Brick or block wall

Small shrubs

Clumps of grass

Sand
Soil
Strong wire netting

An outside run. A house inside the run or attached to it should be available.

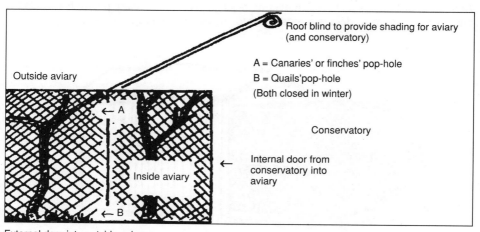

Roof blind to provide shading for aviary
(and conservatory)

A = Canaries' or finches' pop-hole
B = Quails'pop-hole
(Both closed in winter)

Outside aviary

Conservatory

Inside aviary

Internal door from
conservatory into
aviary

External door into outside aviary

A small aviary built onto a conservatory, allowing winter and summer accommodation.

An aviary that can be used for flying birds such as finches and more ground-orientated birds such as quails. The former have access via a top pop-hole while the latter go in at ground level. Photo by courtesy of *Southern Aviaries*.

Aviary systems

Quail breeders and keepers of ornamental or rarer species of quail often use an aviary system, particularly because of the need of some species to perch. The simplest aviary is a small house with attached run, or a bank of these arranged in such a way that the individual runs are parallel with each other, but separate, so that different breeds are kept apart. It is normal for specific breeds to be kept in pairs or trios, but several pairs should not normally be kept together because the males may fight. Hen birds of the same variety can usually be housed together.

There are many purpose-built aviaries available from specialist suppliers, and they can be most attractive, particularly in a garden setting. Many of them are designed for both flying birds and ground orientated ones.

Breeders of the more ornamental varieties like to have runs with natural vegetation. It is possible to arrange this, as long as precautions to deter rats are taken. I have already referred to the fact that concrete runs are the safest, but strong wire mesh laid on the ground and extending beyond the perimeter is a good alternative. Once in position, rubble is placed on top, followed by soil and a layer of sand. Small shrubs or other plants can then be planted directly into the soil. Long grasses are particularly suitable for they emulate the conditions which many breeds of quail frequent in the wild.

As long as there is a quiet, protected corner with such plants, many quail breeds will build nests and incubate their eggs. The Chinese Painted quail, for example, will make a nest in the middle of a clump of long grass, while Coturnix breeds will also become broody, particularly in their second year.

It is necessary to sound a note of warning about the type of netting used underneath the soil in runs such as this. I once used netting which was slightly rusty. I reasoned that it was not good enough for anything else but was quite suitable for putting under the soil. It was a grave error! Rats discovered a weak point and burrowed under the perimeter wall, coming straight up in the centre of the run. They killed the three Japanese quail, one male and two females, who were residents.

A really rat-proof floor is provided by using flag stones instead of netting. These placed close together will exclude even the most devious of rats, while the cracks between them will provide necessary drainage. Rubble, soil and finally sand can then be placed on top as before. Sand is recommended because it does allow the birds to make their own dustbathing depressions, as well as providing a popular scratching area. Some breeders also recommend this as a means of deterring earthworms from coming to the surface. Earthworms are hosts to a number of parasites which can be transmitted to quail.

An outside aviary run also needs protective walls at the base, with wire mesh above. This can be brick, block or wood, and provides wind protection. Within such an aviary any stout house which is rain and draught-proof is suitable. In milder areas of the country, hardier breeds can be left in outdoor runs as long as the house is well insulated.

Chinese Painted quail are frequently kept in aviaries where tree perching birds are kept because they help to clear up the seeds dropped by the flying birds. This should not be regarded as the sole source of food for them however. It is not enough for they need a properly balanced diet like any other species. (See the section on *Feeding*.) In this context, I should mention a lady who kept Java finches, her prime interest. She had introduced several Chinese Painted quail into the bottom of the aviary because someone had told her they were useful to clean up aviary floors. She telephoned me to say that they *"kept dying"* and did I know why? During the course of the conversation it emerged that not only did they have no food other than what the finches left, but had no house or any form of shelter other than the sapling which was planted in the aviary. Yet the Java finches

A small hay-box brooder makes a suitable house within an aviary flight for Chinese Painted quail that are kept as ground companions for flying birds such as finches.

had a proper diet and sheltered accommodation. No wonder they died!

Chinese Painted quail are ground-orientated birds and need to have proper rations and clean water provided for them in suitable containers on the ground, rather than having to rely on the occasional spilt seeds falling like manna from above. They also need proper housing which will provide warmth and shelter, free of damp and draughts. A simple house is all that is necessary, but it needs wood shavings or other warm nesting material to provide insulation.

The movable house and run

If small numbers of Coturnix or Chinese Painted quails are to be kept outside an aviary or house, the best solution is to use a strong house and run of the type used by poultry keepers for hens and chicks. (Bobwhite and larger breeds should be in an aviary where they can perch).

A small house and run can be moved from one area of grass to another, allowing access to clean pasture on a regular basis. The house needs a well boarded floor capable of deterring rats, so that once shut up for the night the birds are safe from harm. The run which has a wire floor resting on the grass, is obviously safer than the floorless type.

With such a system, it is a good idea to keep specific breeds together, rather than mix them up, although females of the various Coturnix varieties will cohabit

A small, movable poultry house and run is suitable for a small number of Coturnix or Chinese Painted quail, although they may not be inclined to use the nest boxes.
Photo by courtesy of *Littleacre Products*.

successfully. Males housed together may fight and should, as a rule, be kept separate from each other. One male can usually accompany up to around half a dozen females, although it is worth remembering that Chinese Painted quail are monogamous and normally kept in pairs.

Remember again, that quail are not naturally hardy birds. While a small house and run may be satisfactory in the warmer months, it may not necessarily be successful in winter. My own practice was to have the birds in a house and run in summer and then to transfer them to indoor breeding cages for the winter. It safer to put the runs on concrete rather than on grass, not only because they are safer from predators, but also to minimise the risk of infection from droppings. They have a box of sand in which to make dust baths, and some branches placed to provide shade and slightly more natural conditions. They do occasionally go on grass however, as the photograph on page 21 shows. It is possible, in milder areas of the country to have quails in outdoor houses and runs in winter, but precautions must be taken to protect them in more severe winters. However, I have heard of a lady in the South of England who has a few quail completely free-ranging in her back garden all through the year. I hope they have a warm house to go into each evening.

Feeding

"So the old woman began to feed the bird with twice as much food as she had given it before; but ere long it grew so fat and sleek that it left off laying any eggs at all."

Aesop's Fables.

Quails need a relatively high protein ration - around 27% when they are growing, reducing to around 20% once they are fully grown. Proprietary chick crumbs are suitable for the young. These normally contain a coccidiostat to protect against coccidiosis, but coccidiostat-free crumbs are also available from specialist suppliers.

Once the birds have reached their productive stage, they need an adult ration. Quail layers' pellets such as those produced by *W. H. Marriage & Sons Ltd* are formulated to meet all the nutritional requirements of laying quail, but without the inclusion of medications. Typical ingredients are: wheat, soya, minerals, fat, limestone grit, vitamins and methionine, and the composition breakdown is as follows:

Oil: 5% Protein: 20% Fibre: 2.7% Ash (mineral content): 11%

The pellets are manufactured to a size of 2.5mm so that they are suitable for intake by birds the size of quail. Around 130g per bird, per week for a Coturnix laying hen is the recommended intake, with fresh water being available at all times. It is possible to make up your own feed, of course, and some suggestions are as follows. Bear in mind that these are not as precisely formulated as a complete ration, and are not recommended for commercial producers.

1 part oatmeal: 1 part chick crumbs: 1 part millet: 1 part shelled canary seed

or *1 part chick crumbs: 1 part canary seed: 1 part millet*

Bear in mind the need to use only chick crumbs without a coccidistat if you are making up a ration in this way. It would obviously not be a good idea to be utilising ingredients with an antibiotic additive on a regular basis, as this can lead to antibiotic tolerance and the encouragement of new, stronger strains of pathogens.

On a small scale, quails enjoy having surplus lettuce leaves, and outer leaves from kitchen garden vegetables. They should also have access to grit, where grain is given, for the proper functioning of the gizzard in breaking down and digesting seeds.

It is worth mentioning that for the chicks of Chinese Painted quail even chick

A range of feeding and drinking equipment suitable for quails

Suspended feeder

Open trough feeder - liable to have feed scratched out of it

Examples of feeders where it is not as easy to scratch out the feed

Hopper-type feeder

Parrot feeder for attaching to cage

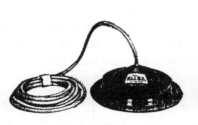

Above: Automatic drinkers for use with a header tank

Below: Range of manually filled drinkers

crumbs may prove to be too coarse for the first few days. I regularly had to grind them up for them, but it was only necessary to do this for about a week.

The most unusual quail diet I ever came across was that adopted by a contributor to *Home Farm* (now *Country Garden & Smallholding*) magazine who lived in a New York apartment. He reared quails in a range of old fish tanks and also raised earthworms and catfish. He mixed his feed in a 30 gallon capacity bucket, using a trowel as a scoop. Here's his formula:

Half a scoop of alfalfa meal: 1 scoop mixed dried roots, seeds & leaves: 1 scoop chopped wheat sprouts: 1 tablespoon iodised salt: 1 tablespoon ground limestone.

This was supplemented with baby catfish, live worms and freshly shredded greens. The New York diet certainly ensures that the quail have enough protein, but there would be few people dedicated enough to follow it.

Feed equipment

Quail are inveterate scratchers and will propel their food in all directions if given the opportunity to do so. Those in commercial cage systems will probably have external feed troughs so that only the head can reach them. Where a floor system is used, suspended feeders which are clear of the ground are the most suitable. If they are gravity-fed or hopper type feeders, the surface area presented to the bird is small enough to prevent jumping up and scratching, yet gravity fills up the feeding area as the feed is eaten.

I have found that using flat open containers is a waste of time. The quail merely jump in and scratch the food everywhere. One might as well put it all on the floor in the first place. If the birds are in a concrete run, and the weather is dry, then adopting this latter course is suitable. A concrete run can be kept clean easily by sweeping and hosing down.

Commercially, feed is available at all times, with the birds helping themselves from the feed troughs on an ad-lib basis. An adult Coturnix quail will consume about 130g of proprietary feed a week. On a smaller scale, feeding twice a day is enough, as long as sufficient feed is given. With my rule-of-thumb approach, I find that one of my handfuls (I have small hands) is enough for six quail at a time, and I feed morning and afternoon. This is fairly general, of course, and there are bound to be variations. Some birds eat more than others, while temperature fluctuations will also have a bearing on consumption: they eat more in winter in order to keep warm. Bear in mind that when it comes to feeding, too much can be just as harmful as too little. An obese bird will be less productive and liable to health problems.

Water

Clean water is essential at all times and there is no question of just making this available a couple of times a day. It must be there whenever the birds feel like drinking, which could be anytime. Again, a commercial unit will normally have an automatic system which utilizes a header tank, tubing, connectors and the drinkers themselves. In a floor system, suspended drinkers as part of an automatic system work well.

On a smaller scale, suspended drinkers which operate on a gravity-principle are satisfactory. This is the type I use for my quail, and I find that refilling once a day is enough to ensure a supply of water for 24 hours. When my quails are brought inside to their wooden canary breeding cages for the winter, I use parrot drinkers. These clip onto the bars of the front of the cage and one of these is enough for a pair of birds for 24 hours. I should add that, in these circumstances, I have no choice but to use flat open dishes for their food, and have to put up with a certain amount of food loss.

Grit

Although a proprietary mixture is said to be a balanced diet without the need of extra grit, I have my reservations about this. It has always been my experience that birds need a small amount of extra grit, and I prefer to err on the side of caution by making it available to them. If grains such as millet, or chopped wheat are given then grit is essential otherwise the gizzard which is responsible for grinding up the grain particles cannot function properly.

On a commercial scale, where feed may be made available on a conveyor system, the grit can be incorporated into the feed at the mixing stage. On a smaller scale, the odd handful given separately on the ground about once a week is sufficient. Fine grit suitable for quail is available from most pet shops or aviary suppliers. Branded supplies usually contain added minerals which help to keep the stock healthy. Grit supplied by poultry suppliers for chickens may prove to be too coarse for quail.

If cuttlefish is provided for beak care, this may also provide sufficient grit and calcium for the digestive system.

Green food

I have already referred to the fact that my quail have kitchen garden surpluses such as overblown lettuce. They quickly shred their way through such leaves with their sharp beaks, and it is arguable that this activity has a beneficial effect in preventing feather pecking or other aggressive behaviour, rather than the strictly nutritional benefits. Most commercial quail would not have access to green food, although it is my belief that quail are healthier, happier and longer-lived when

Male and female Japanese Coturnix quail in an outside run.

they have a good basic diet with lots of variety. If it is true for humans, then why not for quails?

Here are some of the plants, kitchen garden and wild that I have found to be popular with quails:

Sprouted grains (including sunflower seeds and those in canary and finch mixtures)
Lettuce
Parsley (Curled and Hamburgh types)
Spinach (Summer and Winter varieties)
Chicory
Chickweed, *Stellaria media*
Fathen, *Chenopodium album*
Groundsel, *Senecio vulgaris*
Dandelion, *Taxacum officinale*

Wash the plants before hanging therm up for the birds to peck at. They will soon shred them. Obviously, the weeds should come from your own garden, rather than where they may have been subjected to pesticide sprays or vehicle exhaust fumes. You don't have any weeds in your garden! It's not like mine, then.

The Essentials of Incubation

Temperature: 37.5⁰C at centre of egg, reducing to 37⁰C for hatching
Humidity: 45% increasing to 75% for hatching
Turning: 3-5 times daily

Good, healthy and unrelated breeding stock with proven performance

Store fertile eggs at 15-18⁰C at a relative humidity of 75% with blunt ends up, and at angle of 45⁰. Reverse the direction of tilt twice daily. Incubate before 7 days old.

Dip eggs in an egg sanitant and incubate at 37.5⁰C at centre of egg (in a still air incubator this will be 39⁰C when thermometer is 5cm above the eggs). Reduce temperature to 37⁰C two days before hatching.

During the incubating period, the relative humidity is 45%, increasing to 75% for the last two days before hatching

If the incubator does not have an automatic turning facility, turn the eggs 3-5 times a day. Wash your hands before handling them.

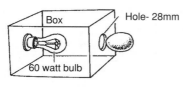

Box — Hole- 28mm
60 watt bulb

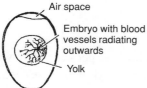

Air space
Embryo with blood vessels radiating outwards
Yolk

After hatching, wait until the chicks are dry and fluffed up before moving to protected brooding conditions with food and water.

Incubation times
Coturnix breeds: 18 days
Bobwhite breeds: 23 days
Chinese Painted 16 days

(These are average times - they may vary by a couple of days each way, and for different breeds).

Candle after 6-7 days if you must to see which eggs are fertile and which are not developing (Markings make candling difficult)

Breeding

"Why should one quarrel with good breeding?"

(Eugene Onegin, 1833)

B reeding is an essential aspect of keeping quail, whether on a large or a small scale. Large production units in the UK tend to breed and rear their own replacement breeding stock, rather than buy in 'point-of-lay' stock, as is the case in the poultry world. The reason for this is because the selective breeding of quail has not taken place to the same degree as it has with poultry. As I have already mentioned, there are some large breeders who selectively breed and sell commercial strains but it is still not a widespread practice. Most producers find that it is better to breed their own replacements, buying in new breeders for the breeding flock as and when necessary.

Smaller enterprises, which often have the coloured and ornamental breeds, generally find that breeding such stock has a ready market among interested poultry keepers, rare breed enthusiasts and aviary owners. Whatever the scale, breeding quail has a fascination which few could doubt. I find it the most interesting aspect of all. The key factor is to obtain good breeding stock to start with, and it is here that a specialist supplier can help. They are listed at the end of the book.

Breeding stock

It goes without saying that the breeding stock should be healthy, have no visible defects and be unrelated. The latter consideration is often overlooked, and purchasers of a pair of a particular breed, may unknowingly have acquired a brother and sister.

Such inbreeding can result in genetic defects being thrown up in the progeny. I once found that a clutch of Japanese Coturnix chicks had three born without claws on the feet. They were the progeny of a pair which I had bought specifically as breeders, although to be fair to the supplier, I did not stress this at the time of purchase. If I had, perhaps the story might have been different. What I should have done, of course, was either to have ensured that they were from different lines, or to have acquired them from different sources. I could also have bought two pairs, one from each source, and swapped over the males.

The use of leg rings is essential in order to keep adequate records and to ensure that you know which bird is which. Normal poultry leg rings are too large, but quail rings are available from specialist suppliers. They can be obtained in different colours, and in a numbered sequence if required. They are available in plastic or aluminium.

Leg Rings

Above: Closed aluminium rings which are put on when the chicks are a few days old. Marked with the current year, they guarantee the age of the bird. Coturnix quails take size R while Chinese Painteds take size L.

Right: Plastic split ring available in 12 single colours or 76 striped colours. They are used for the identification of breeding stock and breeding lines. They can be put on at any age with the aid of an opening tool, as shown. Coturnix quails take size 1FB, while Chinese Painteds take size XB.

Illustrations by courtesy of Southern Aviaries

Before collecting eggs for incubation, the male and females should have been confined together in breeding quarters for at least a week, and it is as well to discard the eggs for incubating for the first few days afterwards (although they should still be collected) until they have settled down into a regular output.

Breeding ratio

Coturnix quail can be kept in pairs, trios or one male for 5 - 6 females. Bobwhites are normally kept in pairs, although there is no reason why the number of females should not be increased. Chinese Painted quail are naturally monogamous and should be kept in pairs. The rarer, ornamental breeds are usually kept in pairs or trios. If you want to keep a precise record of which hen is producing which egg, then the obvious solution is to house only a pair in the breeding quarters. Selective breeding for a particular characteristic is made much easier in this way.

Selective breeding may be for a number of reasons: colour, feather markings, egg production, quick growth, weight - and so on. The basic principle which operates is that if you breed from two birds which both have a similar characteristic, their progeny will tend to have the same feature.

In many commercial units, it is normal to have one male to each four hens in the cages and mixed running is no problem as far as commercial egg sales are concerned. The eggs are collected every morning and graded. The large and the small ones are selected for selling, while the medium-sized ones are retained for incubation. Of those hatched, most will be reared as table birds, while some are retained as future breeders. Stock birds which are eventually selected from these are then housed in trios in separate breeding accommodation.

Female Coturnix showing denuding of the head feathers during the breeding season.

The onset of lay

Coturnix quail will start to lay at around 5-6 weeks old, and the eggs will be fertile from about 6-7 weeks onwards. The behaviour of the male will leave no doubt as to when the fertile stage is reached because he will produce 'foam balls' and deposit them on the ground. Once the male is sexually active, he will mate frequently with the females, gripping the feathers on the top of their heads with his beak. The photograph above shows the head feather loss at this time. Occasionally, the skin is pierced leaving a wound. If this happens, the female should be removed immediately, in case the bloodstain incites aggressive attack. A period in a hospital cage in solitary confinement will soon enable her to recover.

Occasionally, a male will object to a particular female for no apparent reason, attacking her in a vicious way. When this happens, there is no other solution but to split them up permanently. I have tried to 'improve on nature' in this respect, by keeping them together, but nature obviously knows more than I do, because it always ends in failure.

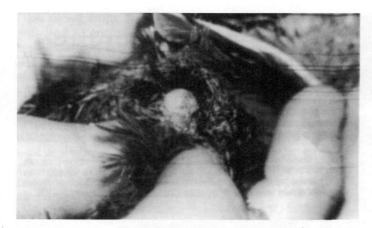

The production of foam balls by the male is a sure sign that he is sexually active.
Photo: Marke Bomer

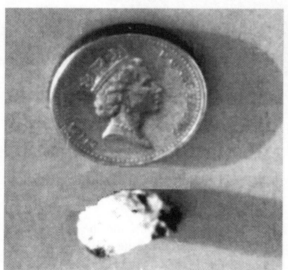

Foam ball deposited on the ground. Compare the size with a pound coin.

Storing eggs prior to incubation

Only the best, undamaged eggs from unrelated breeders should be selected for incubation. Once selected, they should be incubated as soon as possible, and ideally no later than a week after being laid. Some will incubate successfully after this, but hatchability begins to decline after a week. (If you have no choice but to keep them longer, reduce the storage temperature to 12-15°C). Normally, eggs need to be stored in a cool room at 15-18°C with ideally a relative humidity of 75% and with the pointed end down. The photograph on the right shows this operation in progress in the hatchery of a large commercial quail farm.

Quail eggs being set for storage prior to incubation at a commercial hatchery.
Photo by courtesy of *Poultry World*.

The eggs are turned regularly through 45⁰ until they are put in the incubator. On a small scale they can be placed in clean cartons or trays, with one end balanced on a support, while a second support stops it slipping. The carton is then tipped in the other direction, with the change taking place daily. Plastic insert trays from incubators make ideal storage holders, and are easily washed and sterilised before use. In this sense, they are preferable to cardboard egg cartons or trays.

Egg cleaning

One of the major causes of poor hatches is the presence of pathogens which cause disease. Eggs should be collected frequently and with clean hands; thin surgical gloves are favoured by many breeders. If the eggs are slightly dirty they can be brushed clean with a dry nailbrush, but beware of damaging the shell!

Washing the eggs with water to which an egg sanitant has been added is an effective way of minimising disease. Incubator suppliers usually sell sanitants under a number of different brand names. Follow the instructions for a particular brand for their relative strengths can vary. A crucial factor is ensuring that the water is *warm* so that any bacteria on the shell surface are drawn away from the pores. If the water is colder than the egg, the effect can be to draw *in* the bacteria!

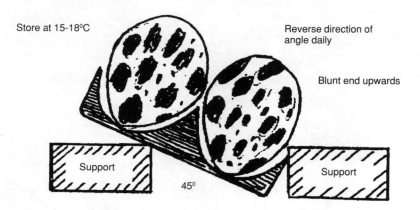

Storing eggs prior to incubation

Store at 15-18°C

Reverse direction of angle daily

Blunt end upwards

Support

Support

45°

Natural incubation

Coturnix quail are not particularly good mothers in captivity, although they will become broody and sit on their own eggs if conditions are acceptable to them. They are more likely to do this in a floor-system or in an aviary, where a certain amount of natural cover and vegetation is present. They can cover about ten eggs at a time.

Bantam hens such as Silkies or Pekins have been used with success to hatch quail, but I have never tried this. In the first issue of this book I asked whether any readers had succeeded in hatching quail by natural methods, and I am pleased to say that I received the following letter about a Pekin bantam who raised 6 Japanese Coturnix quails:

"One of my Pekin bantams (first time mum) raised six quails without any trouble. She sat on seven eggs and treated them as her own chicks. I kept them in a small pen with very small wire, and used a water drinker of the free-standing type which is used for small cage birds. For the feeder I had the base of a drinker which usually holds an upside-down bottle.

Although this is only a one-off experiment, I can't see any reason why it shouldn't work well with any calm type of bantam. I took mum away at three weeks and moved the quails to an aviary at four weeks."

(Elaine P. Samson of St. Andrews in Australia)

After the second edition of the book appeared, I received the following letters about the natural incubation of Chinese Painted and Bobwhite quails:

"One year, a pair of Chinese Painted quails earmarked a sandy corner of the aviary as their own. It was quite by accident that I came across the little nest behind some logs under a shrub. The female was sitting there, motionless, while the male paraded up and down indignantly, doing his best to shoo me away. I crept away and left them to it, disturbing them as little as possible, other than providing food and water nearby.

I was also worried about the finches flying above, so I erected some netting around the little family's area. To my delight, she hatched four little ones, although one died within a few days. I kept a careful eye on the male in case he attacked them but he was as good as gold. I did make sure that the finches were kept clear until the chicks were several weeks old, just in case, but I don't know whether I was being too cautious."

(Jane Hamilton, Surrey, England)

"I don't keep quail, although my grandson has some Pharoahs, but I thought you'd welcome hearing about the Bobwhites that took up in my woodlot. I'd heard their whistles for a while before I found the nest. It was just inside the tree-line, in a little hollow, with the cutest grass roof you could imagine.

I kept away as much as possible, and only saw the sitter for a short while. The little brood hatched when I wasn't looking, too, but I guess they survived - at least until the hunters got them. From the egg shell remains there seem to have been around five chicks. I left some chicken feed and water nearby for them so maybe I gave them a good start."

(E. Hurst, Mississipi, USA)

Two of the author's quails - a British Range and a Bobwhite - cohabiting harmoniously in an outside run.

A small table-top and fan-assisted, incubator with automatic roller turning facilities. The tray at the back is for hatching.
Photograph by courtesy of *Curfew Incubators*

Artificial incubation

The salient points in the incubation and hatching of quail eggs are as follows:

Breed	Temperature			Relative Humidity			Duration	
	Room	Setter	Hatcher	Room	Setter	Hatcher	Pipping	Hatching
Coturnix (Japanese type)	24-28°C	37.5°C	37.0°C	60%	45%	75%	15 days	18 days
Bobwhite	24-28°C	37.5°C	37.0°C	60%	45%	75%	20 days	23 days
Chinese Painted	24-28°C	37.5°C	37.0°C	60%	45%	75%	12 days	16 days

These represent the ideal conditions. In reality, there will be some variation from the optimum, but the closer they are, the better.

Most quails are incubated artificially and certainly on a commercial scale, it is essential. Large units will have a separate setting room and incubation room, or at least a separate setter and hatcher. The first caters for the eggs until they are at the initial 'pipping' or cracking stage. The latter is where the trays of eggs are moved for the actual hatching. On a small scale, the incubator will either have separate shelves for setting and hatching, or at least, a hatching tray in the setting area. On a small scale, most general purpose small incubators are satisfactory for quail. I have used several types of incubator over the years and have usually had reasonable results for quail, poultry and waterfowl with them. It must be said, however, that the rate of hatchability is generally lower than it is for poultry.

There are certain basic elements that are essential to bear in mind for successful incubation. They are shown in the table above.

Ventilation: An adequate flow of air to provide oxygen and disperse carbon dioxide is essential. A still-air incubator relies on the opening and closing of air vents, while a fan-assisted one has a built-in fan to do the job. The latter is certainly a feature worth having if you are thinking of buying a new incubator.

Turning: Regular turning is essential, if the developing embryo is not to stick to one side of the shell membranes, with a resulting malformation of the embryo.

A quail breeder in the West Country, told me that he had consistently poor hatchings with a manually operated incubator, and when he went over to using an incubator with automatic turning facilities, his problems ceased.

If manual turning is necessary, the eggs will need to be turned at least 3 times a day, and ideally 5 times. Place a cross on one side of each egg so that you know which side ought to be facing upwards for each particular turn. The markings on Coturnix eggs can make this difficult, so using a bright colour is a good idea.

A small incubator with automatic turning facilities and adapted for quail eggs.

Newly hatched quails in a small incubator. Photos by courtesy of *Brinsea Incubators*.

Temperature: During the incubation period, the optimum temperature is 37.5°C at the centre of the egg. In a still-air incubator, where there tend to be fluctuations in temperature in different parts of the container, it may need to register as 39.5°C when the thermometer is held 5cm above the eggs, so that 37.5°C is the reading in the centre of the egg. The advice given by the incubator manufacturer is crucial in this respect.

A few days before hatching, the eggs will 'pip'. This is the initial stage when the little birds position themselves to break through the shell membranes into the air space at the blunt end of the egg. From here, the outer shell is then cracked and eventually broken. At this stage, the temperature is reduced slightly to 37.0°C to allow for the fact that the birds are themselves generating heat. Turning of the eggs also ceases at this stage.

Coturnix (Japanese) quail chick hatching. Photo by courtesy of *Poultry World*.

Humidity: The relative humidity indicates the amount of moisture in the atmosphere. During the setting period it needs to be at 45%. This ensures that enough moisture can be lost from the egg, allowing the air space to grow sufficiently for the embryo to breathe. If it is too high (and this is common in northern, damp climates) it results in the 'dead-in-shell' phenomenon. If too low, the bird will be unable to break out of the shell. At the pipping stage, the relative humidity is increased to 75% to compensate for this. This is normally achieved by adding extra water to the incubator or hatcher.

Relative humidity can be measured with a hygrometer. Many small incubators are now equipped with these. They allow the relative humidity to be measured, but do not control it. If it is necessary to reduce humidity, it is a matter of not adding water to the water tray or using a desiccant such as a heated newspaper put in the base of the incubator. Small dehumidifiers which work in conjunction with incubators are also available.

It is frequently forgotten that the siting of an incubator is important, for the outside conditions affect what is happening inside. This is particularly true of relative humidity. I found that when I moved my incubator from the shed into a spare bedroom which had central heating, and therefore less humidity, I had noticeably improved hatches. (In my book, *Incubation: A Guide to Hatching and Rearing* (3rd edition), there is a comprehensive coverage of the whole process of breeding and incubating, including details of how to make your own incubators and brooders).

It is possible to candle, or shine a bright light through the eggs, at around 5-7 days, in order to discover which eggs are fertile and developing, and which are not. However, I have always found that the speckled nature of quails' eggs makes them difficult to candle effectively, so I stopped doing it.

Brooding

It is often claimed that your problems with quails only start once they are hatched. Their small size certainly makes them more effective escapees than poultry chicks. A glance at the photograph on the previous page indicates the size of a newly hatched Coturnix quail. That of the Chinese Painted quail is around the size of a bumble bee!

Any brooder housing will require finely meshed wire to ensure that they do not escape through the holes. I find that for Chinese Painted quail, an old fish tank with netting across the top is the safest brooder. For Coturnix chicks I adapted an indoor rabbit hutch to provide heating from a red bulb. This is confined to what had been the rabbit's sleeping accommodation and the presence of wood shavings as litter provides a warm, comfortable area. (For Chinese Painted quails, sawdust is a better sized litter). Food and water are made available in the living accommodation half.

I had to replace the existing bars of this section with fine wire mesh. That produced by *Twilweld* and available from many large garden centres or aviary suppliers is ideal.

There are, of course, many possibilities when it comes to making home-made brooders and even a large cardboard box can be turned into an effective brooder. Commercial brooding units are also available and, on a large scale, a floor system such as that shown on page 36 is ideal. Here, the quail chicks are on wood shavings in a rat-proof building with heat provided by overhead brooding lamps. Food is made available in shallow containers or chick feeders which allow the head to enter, but prevent them getting in to scratch out the food. Small, gravity-fed drinkers are available and the brooding area is confined to the vicinity of the lamps by wire netting lined with plastic feed sacks as insulation.

Three day old Coturnix quail chick

As the chicks grow and develop, these confining walls are removed and the artificial heat switched off. As a general rule, heat will be necessary for about three weeks, gradually raising the lamps to harden off the young birds. Initially, it should be 37^0C, the same as the hatcher, but this will be decreased every day. If the chicks huddle in the middle, they are cold and the lamp should be lowered. If they are dispersing to the edges, they are too hot, and the lamp should be raised.

The outside temperature naturally has an influence, and in particularly cold periods, it may be necessary to extend the period of heating. Similarly, if it is warm, the availability of artificial heat can be reduced or withdrawn earlier.

Sexing quail

Coturnix laying quail are easy to sex because from about three weeks onwards, the reddish brown chest of the male will start to become noticeable. His markings are also more distinct than those of the female.

A home-made brooder for a small number of young quail.
It could be an adapted rabbit hutch or a large cardboard box.

Section of front cut
away to show interior

Food and water

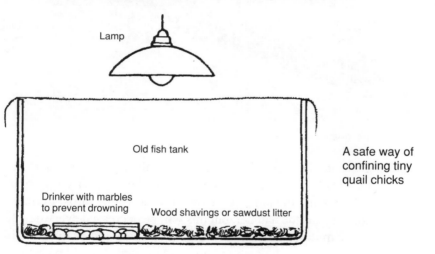

Lamp

Old fish tank

A safe way of
confining tiny
quail chicks

Drinker with marbles
to prevent drowning

Wood shavings or sawdust litter

Once they are adult and in breeding condition, from about 6 weeks onwards, the larger size of the female and the sexual behaviour and foam ball production of the male, referred to earlier, will be apparent.

Before the age of 3 weeks, it is virtually impossible to sex the birds, and it is debatable whether it is necessary at this stage anyway. The same is generally true for other breeds of quail. If there is anyone eagle-eyed and knowledgeable enough to sex the bumble bee sized Chinese Painteds, I shall be interested to know what the secret is!

Dull-emitter heater suitable for brooding young quail.

If the chicks cluster in a ball under the lamp, they are too cold - lower the lamp. If they are ranged around the periphery, they are too hot - raise the lamp.

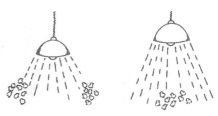

The coloured varieties of Coturnix, such as English White, American Range and Tuxedo do not have an apparent difference in feathering between the male and female. The only way to tell them apart is to go by size difference (the female is bigger), by sexual behaviour patterns and by vent examination.

In the male the vent is more domed, while that of the female is more inverted, as shown on page 67. This difference is onlyseen after they have reached sexual maturity, when their behaviour will indicate their sexes anyway. In adulthood, the difference in vocal sounds, mentioned earlier, is also apparent.

Problems with young quail

Reference has already been made to the fact that young quails are escapees and that all cracks and crevices must be filled. The gauge of any netting used needs to be the finest available.

Their tendency to drown easily in shallow water should also be taken seriously, and for the first week, drinkers need to have clean pebbles or marbles placed in them, so that the depth of water is reduced to a safe level, while still providing water for them to drink.

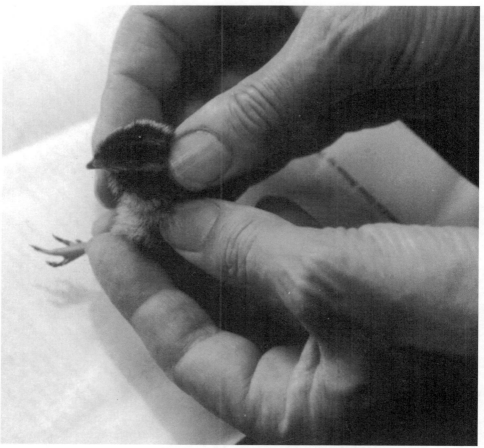

Quail chicks' toes can be damaged by underfloor heating in a brooder.

Chick crumbs may also be too big for them and for the first few days, may need to be crushed.

A problem with feet which I discovered to my cost some years ago was when I used a brooder with under-floor heating. Although equipped with a thermostat, and having adequate insulation as well as wood shavings litter on top, the whole batch of quail developed the same foot problem. Basically this was a blackening of the toes, followed by rapid withering and ultimately the complete loss of them. The whole batch had to be put down and I have never used this form of brooding again. The problem has never recurred.

Chicks born with deformed feet (rather than those subsequently damaged) are usually as a result of gentic defects, from breeding birds that are too closely related, or which have been inadequately fed. The diet may seem to be adequate

An open water container should have pebbles or marbles placed in it to prevent the chicks drowning

but if it is deficient in the right balance of mineral and vitamins, it can show up as defects in the progeny. The following are some of deficiency problems;

Clubbed ends to down feathers - Vitamin B_2 deficiency in parents
Curly toes - Vitamin B_2 deficency in parents
Splayed legs - Vitamin D deficiency in parents
Head thrown back, gazing upwards - Vitamin E deficiency in parents

All of these problems could be avoided with a balanced diet for the breeding birds. If they are not receiving a proprietary feed formulated for quail, it is necessary to ensure that they have a 20% protein intake with their carbohydrate feed, and are given a vitamin and mineral supplement.

Grass ranging and the ability to forage in the sunshine does provide many of the supplements, but few quails have this environment, so the extras must be provided for them. Wheat germ provides Vitamin E to avoid Crazy chick disease (*Encephalamasia*) which causes chicks to 'star-gaze), while yeast, hard-boiled eggs or *Marmite* provide Vitamin B_2. Cod liver oil, or being in the sunshine, provides Vitamin D so that the chicks do not develop Rickets. On a small scale, these elements can be added to the diet quite easily. They should not be necessary, however, if a proprietary feed formulated for quail, is used.

Left: Side view of vent area of male quail
Right: Side view of female vent area

Purpose made cartons are available for quails' eggs.

Female Japanese quail in the long grass environment that would make her feel at home enough to consider sitting on her eggs.

Eggs

"Her mother's cookery book, that edifying work, had been neglected of late for want of the eggs and butter, without which, in its opinion, nothing could be brought into being."

(*Fairies of the Downs*, British Fairy Tale)

The eggs of Coturnix birds are off-white splashed with chocolate brown. The latter may be in the form of tiny spots, large spots or large splodges. There is a considerable variation in patterning. I once tried to keep a record of all the variations of patterning, hoping that I would be able to identify particular eggs as belonging to specific hens, without having to trap-nest. I soon gave up because just when I thought I was getting somewhere, a new variation would appear.

They also occasionally produce olive-coloured eggs with no markings at all, just to be perverse. Occasionally, some strains will produce all-white eggs as a mutation. The size is approximately one third that of a chicken's egg (see the photograph on page 71).

In the wild, about a dozen eggs are laid in a clutch, with two or possibly three clutches in a season. Selectively bred strains can produce between 150-200 eggs without artificial light. The provision of light and good winter management can increase this to 200-300 eggs. It should be emphasized, however, that to achieve such production levels requires the use of good commercial strains, and a high level of management.

In Britain, the same Coturnix quail are generally used for egg production as well as for the table trade. Although there has not been widespread selective breeding and development for commercial purposes, as with poultry, it has occurred on a limited basis. Terry Rolph of *Curfew*, has developed his own 'Crusader' commercial strain, while in the USA, *Marsh Farms* have bred the 'Marsh Pharoah' strains for both eggs and table birds. In Japan, where selective breeding has perhaps been more intensive than anywhere else, various strains of Japanese quail have been developed for eggs, table birds and for laboratory research.

Good commercial strains are therefore available and anyone thinking of starting a quail enterprise, is advised to invest in such stock. Once the enterprise is running smoothly you should also concentrate on doing your own selective breeding, by maintaining a breeding flock from which future quality birds can be selected.

Bobwhite eggs

The eggs are slightly bigger than those of Coturnix breeds, and generally have a paler appearance. The background colour is off-white with a fine speckling of brown spots. There is a variation in patterning, with some eggs having bolder markings than others, but on the whole, they are more finely marked.

In the wild, up to about two dozen eggs are produced in a season. Selectively-bred specimens in protected conditions can produce 50-100 eggs a year, with no extra light. With the provision of artificial light, the provision of winter quarters and adequate feed rations, this can be increased to 150-200 a year.

In the USA, the Bobwhite quail is mainly regraded as a 'managed environment' bird, where tracts of semi-wooded land are managed for the hunting industry. The equivalent, in Britain, is pheasant management. To a limited degree, the Bobwhite is used for egg and table bird production in the USA. In the UK, it is primarily regarded as an aviary bird.

Chinese Painted eggs

The eggs are considerably smaller than those of the Coturnix quail. They are a dull bluish-grey colour. In the wild, about a dozen eggs are produced in a season. In aviary bred and managed birds, this number can be increased to around 50, and the provision of artificial light will further increase the total. However, it is not usual to give Chinese Painted quail artificial light, as is the case with the Coturnix breeds, unless it is to produce early eggs for incubation.

The question is often asked: Can one eat the eggs? There is no reason why they should not be eaten, but they are much smaller than Coturnix eggs, and it seems hardly worth it. There is also the aspect that there is a ready market for Chinese Painted quail within the pet and aviary market, so it is much more profitable to incubate the eggs and sell the young through local pet shops.

Eggs for sale

If quails' eggs are to be produced for sale, then the best choice of breed is a commercial laying strain of Japanese Coturnix. If the enterprise is to be a reasonably large one, sufficient to provide an adequate income, then it is difficult to see how anything other than a cage system or confined floor system, could be viable.

The former is the most efficient in terms of management, as the eggs are easily collected. With a floor system, egg collection is more time-consuming and hazardous. Eggs tend to be laid anywhere and quails do not have a well-developed nest-laying instinct as do chickens. I have frequently stepped on eggs laid on the floor. They are small and the speckled nature of the shell has a camouflaging effect against the floor litter. In Britain, most quails' eggs are produced in cage systems while table birds are produced either in cage or floor systems.

Left to right: Chicken egg, Coturnix egg, Chinese Painted egg.

Eggs should be collected at regular intervals on a daily basis and stored in a cool room at around 15⁰C. If any are to be incubated, it is best to select medium sized ones for this purpose, leaving the larger and smaller ones for sale. At present, there is no official system of grading the eggs, as is the case with chickens' eggs.

With a cage system, it is unlikely that there will be any problem of soiling of eggs. If there are any lightly soiled eggs it is best to brush them with a clean, dry nailbrush. Heavily soiled ones can be washed in warm water with a purpose-made egg sanitising solution, but it is best to avoid selling soiled eggs. If they are cleaned with a sanitising solution, they are suitable for incubation purposes.

Packing and selling eggs

Eggs are sold in several ways. The fresh eggs frequently go to hotels, quality restaurants, delicatessens, farm-gate shops or other retail outlets. Packaging is an important aspect of marketing and plastic egg cartons specifically for quail eggs are available from specialist suppliers. If the quantity of cartons purchased is large enough, it is usually possible to have descriptive labels with details of your own farm incorporated. Alternatively, it is possible to order labels that you can stick on yourself. A company such as *Danro* can provide for all the labelling requirements.

Preserved eggs

A common form of egg marketing is to hard boil the eggs and put them in brine in a glass jar. The example shown left has six eggs in a jar with a metal crimped-on lid. This particular example is a product imported into the UK from France. Alternatively, cooked, shelled eggs in brine or pickling vinegar are possibilities. Quail eggs can also be used to produce a prepared product such as miniature Scotch eggs. (Recipes are given later).

Winter eggs

The provision of artificial light in winter is essential if egg production is to continue without disruption. In a small house, a 25 watt bulb is adequate for about a dozen birds and 40 watt for around 50. A 150 watt bulb with a reflector will be sufficient for a 9 x 4.5m (20' x 10') building. Fluorescent tube lighting is also satisfactory. It is essential to incorporate a time-switch in the circuit so that the light is switched on and off automatically. A total of 15-16 hours of light a day (natural plus artificial) will keep egg production at satisfactory levels.

Start giving extra light before dawn or dusk (or a combination of both), gradually increasing the duration to compensate for the shortening days. Similarly, as the days lengthen, gradually decrease the amount of artificial light.

It is important to be consistent in turning on and switching off the light, and it is here that a pre-programmed time-switch is so valuable. In winter, the drop in temperature imposes more demands on the metabolism so that a certain proportion of energy is needed to keep warm. Bear this in mind and if necessary, increase the rations accordingly.

An important aspect to bear in mind is that giving extra light pre-supposes that the birds are all females. If they have not been separated, the males will fight, and extra light will only aggravate this. If a mixed group of birds are being reared for the table at a relatively early stage, where separation would not be feasible, it is best to restrict light.

Because of their small size, quails can be fiddly to truss, but with practice it becomes easier.

Quails for the table

In Britain, the same strains of Coturnix quail are used for egg production as well as for the table. In the USA, the Bobwhite is sometimes raised for meat.

Commercial battery systems and floor systems are both used for rearing table birds. A house with a concrete floor is the most appropriate because of the dangers associated with rats which can kill adult quails with ease.

Ridge ventilation in association with side windows or vents is ideal, so that ventilation can be controlled. A very large house will probably rely on fan ventilation, but this should not be necessary if the tendency to overstock is resisted.

Many producers find that dividing a house into separate pens is better than having one large house. Quails are nervous birds and the tendency to panic and flap in one direction can lead to problems of management in this respect. An easy and relatively cheap way of subdividing the floor space is to use game netting, or some other form of lightweight panelling. Most game equipment suppliers sell such products.

Each pen or floor space needs wood shavings or sawdust litter, a suspended or gravity-fed feeder (or one from which the birds cannot scratch food) and ideally, an automatic watering system. Young quail will have artificial heat as described in the breeding section.

A certain proportion, or indeed all the females, may be separated for future egg production, leaving the males behind for table production. The aim will be to produce quail with a liveweight of 160-200g (6-7oz) at the age of 6 weeks, with a conversion ratio of around 3:1. In other words, for every three ounces of food consumed, around one ounce of weight is produced.

It is not feasible to weigh all the birds, of course, but it is a good idea to take a few sample weighings, using say half a dozen birds. This will give a general indication of the average weight of the batch.

The easiest way to weigh a quail is to use a small bag with a draw string at the top. The bird is popped into the bag, the draw strings tightened and then the whole thing is suspended on a spring balance. Those made by *Salter* are ideal. It takes only a moment or two, there is no distress caused to the bird, and it is released immediately after weighing. Ready-made weighing cones and scales, as well as electronic scales, are available from specialist suppliers.

Weighings

Take sample weighings of about 6 birds per batch in order to have an average weight every week until killing time.

Keeping records

Everyone has their own system of keeping records. The form is not important. What is necessary is to ensure that all details in relation to weight gain and feed consumed are recorded.

Date hatched _ _ _ Number _ _ _ _	Average Liveweight	Feed consumed	Comments
Week 1			
Week 2			
Week 3			
Week 4			
Week 5			
Week 6			
Week 7			
Total			Losses _ _ _ _

Light restriction

Table quail grow more quickly and sexual development is slowed down if the amount of light to them is restricted. This does not mean that they have to be kept in a twilight zone, as is the case with so many poultry broilers, but no artificial light should be made available to them, as for egg producers. Some producers, I am afraid, go to absurd lengths in blocking out the light reaching the cages. I am against this on humanitarian grounds. Every creature has a right to natural light! The aim is to cut down bright light, while allowing a certain amount of natural light to come in. This suppresses the urge to mate and to lay eggs so that growth continues uninterruptedly. In this way, much heavier weights are gained.

Marsh Farms in the USA claim that delaying the onset of lay until 12 weeks of age produces a doubling of normal bodyweight.

Killing, plucking and dressing

The usual method of killing is to sever the head in one quick movement. For this, a sharp butcher's knife or cleaver used in conjunction with a wooden block is suitable, although some producers find that sharp shears are effective. Alternatively, poultry equipment suppliers sell a purpose-made poultry killer which operates on a guillotine principle.

Allow the birds to bleed, but it is not necessary to hang quails as it is with other game birds. Plucking should take place as quickly as possible, and here there are several alternatives:

74

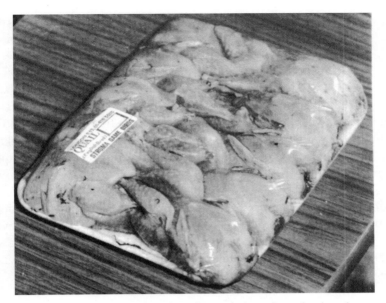

Plucked, gutted and packaged quails ready for sale as fresh or frozen.
Photograph by courtesy of *Poultry World*

Dry plucking: This is simply the removal of the feathers without using water or any other medium to assist the process. Plucking by hand can be a rapid process for those experienced in the field, but mechanical dry plucking machines are available where large quantities are involved. Many people use these for the rough plucking, finishing off the pin feathers by hand, scraping or using a wax finish.

Wet plucking: Here, the bled birds are immersed in scalding water for a moment then removed and plucked as soon as possible. When plucking is complete, they are immersed in chilled water to cool.

Waxing: The principle of waxing is that the birds are dipped in molten wax and as this cools and hardens, the feathers come off with the wax crust. As referred to earlier, some producers use this method as a 'finishing' technique after initial rough plucking.

Once plucked, the birds are drawn. Traditionally, the innards were not removed and Mrs. Beeton would have frowned on such a practice. Modern susceptibilities however, demand that gutting is as complete as possible and commercial table quails will certainly need to be processed in this way, unless they are being sold by contract to a butcher for processing. Gutting is not as easy as it is with poultry, bearing in mind the relative smallness of the birds. Utilising a spoon inserted in the neck end and rotating it inside the body cavity is effective. Cutting around the vent and enlarging the opening then allows the innards to be drawn out.

Some producers find this process too time-consuming, and slit the birds down the back in order to gut them. Some complete the process by boning the birds at the same time, selling the finished product as 'boned quail'. On a large scale, vacuum drawing of the innards is carried out. Once gutted, with the neck cut off close to the body and the legs cut off at the first joint up from the foot, the birds should be chilled again while awaiting packaging.

The usual way of packing quail is on polystyrene tray containers and covered with cling film sealed into position. A common procedure is to package four birds to each tray, arranged neatly, breast-side upwards. The photograph on the previous page shows ten birds packaged in this way, with a label showing the producer's name and the description that they are 'English bred quail'.

Oven-ready birds, as described above, can be sold fresh or frozen, depending on the particular market requirements. One great advantage of quail over game birds is that they are available right through the year, rather than on a seasonal basis. This is an important marketing point, in making 'game' dishes available to connoisseurs all through the year.

Smoked quail

A popular delicatessen commodity, smoked quail, is definitely for the top end of the market. If an enterprise is to be expanded in this direction, the best way is to arrange a contract with a local smoking company so that they do it in batches as necessary. Smoking of food is a skilled task and as there are EU directives with regard to safety, it is better to let a specialist company do it.

The recipes at the end give details of cooking quail to good effect. When selling direct to customers, as for example in a farm shop, being able to give them advice on how to cook them is important if they have not tried quail before.

Quail meat regulations

In order to sell quails for meat it is necessary to be familiar with the 'meat hygiene' regulations. These vary according to the scale of operations. The regulations governing quails shadow fairly closely those for poultry.

The starting point for information is to contact the *Veterinary Meat Hygiene Advisor* at the local office of the *Ministry of Agriculture, Fisheries and Food*, who will explain what regulations are applicable in your particular case.

The regulations cover the slaughter and sale of fresh or frozen quail meat. Small scale operations, slaughtering less than 10,000 birds a year, are exempt from *Directive 92/116*, but are restricted to selling in their own localities and traditional markets. If you plan a larger operation, producing more than 10,000 but less than 150,000 birds, you qualify as a 'low throughput plant'. The more stringent regulations covering this, and larger scale operations, are available from the *Veterinary Meat Hygiene Advisor*. Finally, if you plan to export quails or quail meat, there are different regulations contained in the *Animal Health Circular 92/137* and *European Directive on Farmed Game 91/495*. Again, the local *Veterinary Meat Hygiene Advisor* can advise.

Quail egg regulations

Regulations affecting the sale of quail's eggs are similar to those appertaining to chicken eggs. On a small scale, the sale of occasional, surplus eggs do not come under the regulations at all. Once more than 250 *breeding* birds are kept for the production of *laying birds for sale*, then the site must be registered and birds tested for salmonella on a regular basis.

Eggs may be sold direct to the public, at the farm-gate, market, or door-to-door. There is no official system of grading quail eggs by size and quality.

If production is to be on a large scale, consult the *Regional Egg Marketing Inspector* for the area. The address is in the local *Yellow Pages*.

Marketing

Marketing is always the difficult aspect of any enterprise, particularly where the production side may take up most of the time. The main demand is for quail eggs and table quail for the delicatessen market, although the potential sales of breeding stock should not be overlooked. The sales of ornamental breeds of quail can also provide a valuable side-line.

Local delicatessens, butchers, game dealers and hotels are the obvious places to start. In fact starting on a small scale has much to recommend it. A farm shop on site is worth considering as long as the site is in reasonable proximity to an urban or suburban area. Planning permission is usually required for 'change of use' of premises, as well as for the setting up of a new shop building.

Small local sales are also possible through local markets. If distribution further afield becomes a possibility, and here production must be regular and consistent, it is only possible on a relatively large scale. A distributor will also be required. In recent years, the supermarket chains have shown an increasing interest in quail products.

Specializing and turning the product into a 'value-added' one will give it even more distinctive value. Reference has already been made to the production of luxury items such as pickled eggs.

An aspect which is not often considered is to open part of the enterprise to the public, particularly if near an urban area. A site where different types of quail can be seen in aviary conditions provides a source of interest. Guided tours of the commercial production side (as long as this is not factory-farming) can also provide a useful and educational experience for school parties.

On the question of diversification, *ADAS* (*Agricultural Development Advisory Service*) can provide help, advice and information, although consultancy services will be charged. The address of your nearest *ADAS* office will be in the local telephone directory or *Yellow Pages*.

Finance

For those intending to keep quail as a small business venture selling eggs, meat or breeding stock, there is the question of finance. Fortunately, setting up a quail unit is not very expensive, and it can be done on a step-by-step basis, expanding to meet demand. There are only a few big, full-time quail enterprises in Britain. The others are on a smaller scale and are run on a part-time basis. There is nothing wrong with this. In fact, making use of empty farm buildings and sheds for a part-time enterprise such as this makes good economic sense. It is often a number of smaller enterprises which contribute to the success of an overall lifestyle.

As the aim is to make a profit, a detailed knowledge of all costs is required. This will enable you to assess how to price the end products.

If the borrowing of money for expansion is required, the best place to begin is the bank. The manager will want to see ideas in figures showing projected costs and sales. The very act of preparing such a plan will be useful and the bank manager's advice and experience should be helpful. There are many useful guides to starting a business in book shops, as well as free ones in some local banks.

Do not neglect proper insurance covering your livestock, buildings, equipment, transport, etc. There are any number of companies who will provide appropriate cover and a good insurance broker should be able to help you.

Costs

Costs consist of *fixed costs* and *variable costs*. Fixed costs are large purchases like buildings, cages, feeders, drinkers and things that last over a period of time. Fixed costs can spread over a period at so much per month, according to your estimate of their life and depreciation. Variable costs or running costs include everything else, such as feed, transport, stock, heating, etc - in fact everything which can be regarded as a cost. Don't forget that everything that you use, particularly your car, can be assessed as a cost.

Keep careful note of all costs in writing, including copies of all invoices. Have a ledger showing sales and costs and keep it on a regular basis. You can obtain advice from your bank manager or accountant. It may be easier to employ a part-time book-keeper to keep everything in good order for a few hours work each month. If you use a computer, there are some excellent software packages around which are specifically geared to small businesses, and which will do everything, including VAT calculations. A good example is *QuickBooks*.

Health

Quails, like all living things, require good food, clean water, warm, dry and draught-free housing and regular attention. The main priority is always to prevent trouble and ill health, and to develop a ready eye for problems at an early stage. As soon as anything suspicious is noticed, such as listlessness, poor appetite, discharge from the beak or unusually coloured droppings, it is a good idea to isolate the bird immediately. This does two things; it stops the spread of possible infection to the others, and it allows the invalid a better chance of recovery.

Listless birds invite aggression and may be severely pecked by the others.

A hospital cage

A hospital cage is a useful thing to have. This could be a wooden bird breeding cage with a wire front, or even a large cardboard box. Specialist suppliers also sell purpose-made ones. Place some woodshavings in the bottom and clip on a feeder and drinker. The chances are that a sick bird will not be interested in the food, but it will certainly require water.

The cage should be in a warm, sheltered area, but if it is particularly cold, it may be a good idea to have a dull emitter bulb fixed up to provide a source of warmth. This can be placed outside the cage so that it is shining in sideways, through the cage front. Depending on the facilities available, it could be suspended from above. It is surprising how the provision of warmth and cosseting in this way, can sometimes make the difference between survival and loss, when all else possible has been done.

It is obviously not economic to call in the vet to see a single bird, although there is nothing to prevent you taking a container with bird along to the vet.

Where the condition is a minor one, such as a cold or simple digestive upset, it will clear up of its own accord, and the protected conditions help to make this sooner rather than later. If it is a more serious condition, there is not a lot to be done, and at least the bird will have met its end in relatively quiet comfort. If the bird is the first casualty in a more general outbreak, it enables action to be taken in time to protect the others.

Commercially, a record of all medicinal treatments is required to be kept. It makes sense to do so on a small scale, too.

Bacterial respiratory infections

Illnesses caused by bacteria can be treated with antibiotics, where viral ones can not. Some of the more common bacterial infections which affect the respiratory tract are bronchitis, pneumonia, infectious sinusitis and chronic respiratory disease. They all have similar symptoms of wheezing, laboured breathing, nasal discharge and loss of appetite.

Antibiotics are available only from veterinary surgeons. They should be used only when it is advisable to do so, and following the recommended doses. If there is a withdrawal period, where eggs or birds may not be sold or eaten, it is essential to follow the guidelines.

Bacterial digestive infections

Enteritis results in greenish droppings. Coccidiosis produces greenish droppings but of a more slimy nature. Antibiotic treatment is required and the vet will prescribe an appropriate one for dosing the drinking water.

Salmonella is always a danger, and every effort should be made to avoid it. Make sure that all feeders and drinkers are cleaned regularly and thoroughly, and that feedstuffs are fresh and clean.

Worms

Internal worms may be a problem with quail which are aviary housed, or which have access to the soil via pens. The importance of moving pens to fresh, clean ground on a regular basis is obvious, as well as the practice of liming aviary soil each autumn while the birds are in their winter quarters.

Caged quails are unlikely to have worm problems because they are not in a position to ingest the cysts which produce them.

Internal worms cause emaciation and loss of feather condition, and any permanent quail breeding stock which has access to the ground, should be wormed as a matter of course before being brought in for the winter. There is only one vermifuge which is licensed for use with gamefowl and poultry, and that is *Flubenvet*. It is administered in the feed and there is a withdrawal period.

Scaly leg

This is a condition of the legs, where burrowing mites push up the scales of the legs and produce white encrustations. It is highly infectious and will spread rapidly to all the birds if not treated quickly. Using an old toothbrush and warm soapy water, soak and brush off the encrustations, then dry the legs before applying a mite killer which is available from the vet or licensed supplier. This is most effective and far better than the traditional paraffin treatment. Do not try and pull the dry encrustations away otherwise the skin itself is pulled away.

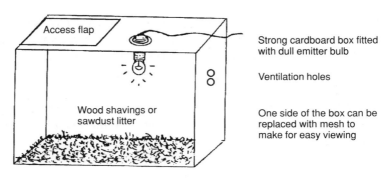

Access flap

Strong cardboard box fitted with dull emitter bulb

Ventilation holes

Wood shavings or sawdust litter

One side of the box can be replaced with mesh to make for easy viewing

A home-made hospital cage

Bumblefoot

A hard lump can sometimes form underneath the foot, where a small wound may have healed over leaving some infection behind. It is first detected when a bird is seen to be limping. If the infection is still active, the foot will feel hot and swollen and antibiotic treatment may be needed. If the lump is near the surface, it can be lanced to remove the pus, then treated with any disinfectant cream.

Wounds

Wounds such as those acquired by head banging or fighting should be cleaned and treated with antiseptic cream or disinfectant. Keep the bird in isolation until the wound has healed, in case it is pecked by the others.

Mites

Quails can be affected by mites in the same way that poultry beome infested. Red mite is a particular problem. Proprietary insecticide powder or spray is the answer, applied under the wings and around the neck and rump feathers. It is available from licensed poultry and pet suppliers. Don't forget to treat the housing and dustbath areas as well!

Breeding problems

Reference has already been made to the necessity of having good breeding stock, and not to breed closely related strains unless you know that they are free of defects. If the introduction of females to males is carried out carefully, there is usually no problem, although it does occasionally happen that a male will refuse to accept a female. I once had a Coturnix layer who was constantly driven off by the male (she was part of a trio), and he devoted his attentions to the other one. He would not mate with her and pecked her unmercifully. I eventually had to take her and put her in solitary confinement. Her wounds healed, she seemed fit and healthy, with a good appetite, when without warning, she died. I have often wondered whether the male knew more than I did, and whether there is some selective survival mechanism which operates with discrimination in these cases.

An aspect which is frequently forgotten is that breeding birds do need an excellent and well-balanced diet, otherwise nutritional deficiencies may manifest in weakly chicks. Clubbed feathers and leg problems in the chicks are associated with an inadequate level of minerals and vitamins in the parent stock. (For details, see page 67).

Egg binding can sometimes be a problem. This is where a female occasionally is seen to sit for long periods of time without laying an egg. The best solution, as with other poultry, is to hold her over steam (a bowl of hot water - but take care not to drop her), and gently massage the vent with *Vaseline*. This is really all that is possible for if the egg breaks inside her, it invariably leads to death. Although infection is usually given as the reason why death occurs, it seems to me that shock plays a major part, for a bird can die before infection has had a chance to develop.

I have already referred to possible problems with young quail. If a good level of hygiene and incubation practice is maintained, problems will be minimised. The incubator and brooder area needs complete cleaning and disinfecting after use and before a new batch of eggs and young are introduced. Most problems with poor hatches are usually connected with one of the following reasons: poor breeding stock, low fertility of parents, poor standard of nutrition in parents, infection in eggs caused by poor standards of hygiene, temperature fluctuations, humidity fluctuations or inadequate turning of the eggs.

Feather and vent pecking, and cannibalism

These are all variations of the same problem, and the causes are many and varied. Check that there is no mite problem, and that feeding and watering levels are adequate. Over-crowding and lack of interest in the environment are often contributory agents. Feeding greens helps to reduce boredom, as well as providing a wider range of nutrients. A bird which is being pecked regularly, or one that is a persistent attacker, should be separated for a time so that the habit can be broken.

Notifiable diseases

Finally, quail can be affected by Newcastle disease or Fowl pest, although it is primarily found in chickens. Symptoms are paralysis of the legs and throwing the head backwards. It is a notifiable disease in Britain so the vet or local office of the *Ministry of Agriculture* should be notified. Infected flocks are normally slaughtered by the authorities, and compensation is available.

Recipes

"The best thing about recipes is that you don't have to follow them."

Just after I wrote the first edition of this book, and my head was still full of quail facts and figures, I went to northern Spain with my husband. On our first evening there we went out for a special meal and I ordered what appeared to be an exotic dish that I had never tried before. When it arrived, it was quail! *"Oh, no!"*, was my first reaction, but as it happens, the meal turned out to be delicious.

It is not necessary to hang quails, as it is with other game birds, merely to allow them to bleed after the heads are removed. Traditionally, the innards were not removed and Mrs. Beeton would have frowned on such a practice. Modern susceptibilities, however, demand that gutting is as complete as possible.

Mrs Beeton's recipe
Ingredients: Quails, butter, toast.
Method: These birds keep good for several days and should be roasted without drawing (*but not if you have modern susceptibilities - author*). Truss them with the legs close to the body and the feet pressing upon the thighs. Place some slice of toast in the dripping pan, allowing a piece of toast for each bird. Roast for 15-20 minutes; keep them well basted and serve on the toast.

Quick roast quail
Ingredients: Quails, 2 slices of bacon per bird, seasoning
Method: This is my favourite way of cooking them and the bacon adds flavour, as well as preventing too much drying. Wrap up each bird in the bacon and use wooden sausage sticks to keep them in place. Roast in a medium oven for about 20 minutes and serve with any of the sauces normally used with chicken or game.

Grilled quails
Ingredients: Quails, butter, lemon juice, bacon slices, seasoning, breadcrumbs.
Method: Slit each quail down the backbone. Flatten each side and sprinkle with lemon juice, salt and pepper. Dip in melted butter and roll in breadcrumbs. Grill for about 5-6 minutes on each side. Meanwhile, roll up the bacon slices and grill them for the last few minutes with the quail. Serve quails with bacon rolls and garnish with mushrooms.

French roast quail

Ingredients: Quails, butter, vine leaves, toast, seasoning, milk.
Method: Wipe quail inside and out. Place in a saucepan with just enough milk to cover them. Simmer gently for 6-7 minutes. Remove and put the saucepan and its contents to one side for the moment. Smear the quail with butter and wrap in vine leaves. Sprinkle with salt and pepper and place in a buttered oven-proof dish. Roast for 10 minutes in a medium oven, then take the dish out of the oven. Strain the contents of the saucepan and pour over the quail. Replace the dish in the oven and cook for another 10 minutes. Serve each quail on a slice of toast and garnish with wedges of lemon and sprigs of parsley.

Miniature Scotch eggs

An enterprising British quail egg producer has evolved a novel way of giving his eggs value-added appeal. He hardboils them, removes the shells, coats them in flour and sausagemeat, with a final coating of breadcrumbs. Once deep-fried for a few moments, they provide miniature Scotch eggs, a delicacy which has obvious appeal for the delicatessen market.

Pickled eggs

Quails' eggs are so attractive that many people hard boil and serve them in their shells. An alternative is to preserve or pickle them in one of the following ways:

Whole eggs in brine: This is a compromise where the eggs are first hard boiled and then placed in a brine solution without removing the shells. A normal brine solution of 50g salt per 600ml of water (2oz per pint) is suitable. Hard boil the eggs, drain and place them in sterilised jars and top up with the brine solution, making sure that they are completely covered. (Jars can be sterilised by washing in hot water, then put upside down on a rack in a warm oven, with the door open).

Home Farm pickled eggs: This is the recipe that I always used for my quails' eggs. Hard boil the eggs, then plunge into cold water to cool and make them easier to peel. (Note that fresh eggs are more difficult to peel than older ones, so don't use fresh eggs for pickling).

Peel the eggs then place them into sterilised glass jars and top up with a pickling solution made as follows: Place 1 litre white wine or cider vinegar and half a litre of water in a stainless steel pan. (You can leave out or reduce the water if you like a 'tangy' egg, but it does mean that the more delicate egg is dominated by the vinegar). If you have just a few eggs, reduce the ingredient amounts accordingly. Add 2 level tablespoons of salt, a medium sized chopped onion and a sachet of pickling spice, and bring to the boil. Simmer for a few minutes, leave to cool and strain. Pour over the eggs, then pop in some of the boiled peppercorns, just to make the jar look more interesting.

Oak Ridge pickled eggs: This is a recipe from Max and Mary Crawley of Arkansas, USA, and I am grateful to Edmund Hoffman for the information.

Boil the eggs for five minutes in water to which the following have been added: half a cup of salt and 1 fluid oz of vinegar per gallon of water. Cool in cold water and peel the eggs, then place them in sterilised jars.

Prepare the pickling solution as follows: add 1 cup of salt and 1 large chopped onion (with other spices if liked) to two pints of white vinegar and two pints of water. Bring to the boil, drain and pour over the eggs.

Large scale pickled eggs: This recipe is more appropriate to large producers. Hard boil the eggs and then place them in commercial grade, distilled white vinegar for 12 hours. This dissolves the shells, but leaves the membranes intact. Wash the eggs and refrigerate until quite cold then immerse in clean water while the membranes are removed by hand. They are now ready for immersing in the pickling solution of your choice.

Quails' eggs in aspic

Quails' eggs in aspic is, of course, a well-known delicacy and another way of achieving the maximum return on the eggs if they are to be sold. Aspic is available in powder form for mixing as required.

The eggs are hard boiled, then shelled. They can be cracked, placed in cold water and have the shells and membranes removed simultaneously. If cut in half and placed in a small glass or plastic container with the yolk facing upwards, the aspic mixture can then be poured on. Once set and cooled, clear snap-on lids can be placed on the container, enabling the product to be seen. If the container is small containing, for example, four halves, it is a perfect individual hors d'oeuvres, and can be marketed as such.

Further information

Books
QUAIL: THEIR BREEDING & MANAGEMENT. G.E.S. Robbins. World Pheasant Association. 1984. (UK)
DOMESTIC QUAIL FOR HOBBY & PROFIT. Robbins. A. B. Incubators. 1989. (UK)
QUAIL MANUAL. Albert F. Marsh. Marsh Farm Publications. 1976. (USA)
RAISING BOBWHITE QUAIL FOR COMMERCIAL USE. Circular 514. Clemson University & U.S. Department of Agriculture. 1964. (USA)
THAT QUAIL, ROBERT. Margaret Stranger. J.B. Lippincott Co. 1966. (USA). Hodder & Stoughton 1967 (UK).
THE ATLAS OF QUAILS. David Alderton. T.F.H. Publications. 1992 (USA)
COTURNIX QUAIL. Edmund Hoffmann. Self-published. 1988. (Canada)
THE CHINESE PAINTED QUAIL. Leland B Hayes. 1992. (USA)
THE CARE, BREEDING & GENETICS OF BUTTON QUAIL. G P Landry. Acadiana Aviaries. 1998. (USA)
UPLAND GAME BIRDS. Leland B Hayes. 1997 (USA)
INCUBATION: A GUIDE TO HATCHING & REARING. (3rd edition). Katie Thear. Broad Leys Publishing. 1997. (UK)

Magazines
COUNTRY GARDEN & SMALLHOLDING, Station Road, Newport, Saffron Walden, Essex CB11 3PL. UK. Tel: 01799 540922. Fax: 01799 541367. E-mail: cgs@broadleys.com
POULTRY WORLD, Carew House, Wallington, Surrey, SM6 0DX.
CAGE & AVIARY BIRDS, IPC, Kings Reach Tower, Stamford St, London SE99 0BB

Suppliers of quail breeds
CURFEW COTURNIX QUAIL, Southminster Road, Althorne, Essex CM3 6EN. Tel: 01621 741923. Fax: 01621 742680. E-mail: incubate@curfew.co.uk (*Crusader* strain of commercial Coturnix quail)
GAME FOR ANYTHING, Rosehays, Clayhidon, Devon EX15 3TT. Tel: 01823 680092. (Japanese, Chinese Painted, Blue Scaled, Rain, Californian, Mutation Coloured Bobwhite, Gambel)
GRANGE AVIARIES & PET CENTRE, Hillier Garden Centre, Woodhouse Lane, Botley, Southampton, SO30 2EZ. Tel: 01489 781260. Fax: 01489 786542. (Chinese Painted, Californian, Bobwhite and Coturnix)
MEADOW VIEW QUAIL, Meadow View Farm, Church Lane, Whixhall, Shropshire, SY13 2NA. Tel: 01948 880300. (Japanese quail in various colours)
NEWMARKET AVIARIES, Newmarket Lane, Clay Lane, Clay Cross, Chesterfield, Derbyshire. Tel: 01246 863506. (Bobwhite, Mountain, Mexican Speckled, Mexican Blue-scaled, Gambel, Californian, Chinese Painted in various colours, including White, Gold, Silver, Normal Pearled and Pied, Japanese quail in various colours)
QUAIL WORLD, Glenydd, Penrhiwllan, Llandysul, Dyfed, SA44 5NR. Tel: 01559 370105. (Japanese quail in various colours)

Equipment and Supplies
ASD COURT STEEL, Thames Wharf, Dock Road, London E16 1AF. Tel: 0171 476 0444. Fax: 0171 476 0239. *(Weldmesh)*

ASH & LACY MONCASTERS, Belvoir Way, Fairfield Industrial Estate, Louth, Lincs, LN11 0JG. Tel: 01507 600666. Fax: 01507 600499 *(Aviary meshes)*

BRINSEA PRODUCTS LTD, Station Road, Sandford, North Somerset, BS25 5RA. Tel: 01934 823039. Fax: 01934 820250 *(Incubators and brooders)*

CURFEW INCUBATORS, Southminster Road, Althorne, Essex CM3 6EN. Tel: 01621 741923. Fax: 01621 742680. E-mail: incubate@curfew.co.uk *(Incubators, quail cages, brooders, header tanks, drinking systems, feeders)*

DANRO LTD, Unit 68, Jaydon Industrial Estate, Station Road, Earl Shilton, Leicester, LE9 7GA. Tel: 01455 847061/2. Fax: 01455 841272. *(Labels for egg boxes and produce, hand-held labellers, sealing tape and cartons)*

ECOSTAT, Tesla House, Tregonnie Industrial Estate, Falmouth, Cornwall, TR11 4SN. Tel: 01326 378654. Fax: 01326 378069. *(Incubators, brooders, DIY incubator kits, digital thermometers, incubator spares and conversion kits)*

GAMEKEEPA FEEDS LTD, Southerly Park, Binton, Stratford-upon-Avon, Warks. CV37 9TU. Tel: 01789 772429. Fax: 01789 490536. *(Quail feeds, incubators and equipment)*

GRANGE AVIARIES & PET CENTRE, Hillier Garden Centre, Woodhouse Lane, Botley, Southampton, SO30 2EZ. Tel: 01489 781260. Fax: 01489 786542. *(Aviaries, aviary panels, mesh, cages)*

INCUBATOR SPECIALISTS, 71 Carlton Avenue, Worksop, Notts, S81 7JY. Tel: Tel: 01909 470800 *(Incubators, brooders and other equipment)*

MANOR FARM GRANARIES, Brington, Huntingdon, Cambs, PE18 0PY. Tel: 01832 710235. Fax: 01832 710326. *('Quail mix' feed, incubators, cages and other equipment)*

W. & H. MARRIAGE & SONS LTD, Chelmer Mill, New Street, Chelmsford, Essex, CM1 1PN. Tel: 01245 354455. Fax: 01245 261492. *(Quail layer's pellets)*

MS INCUBATORS, Olney Park Cottage, Yardley Road, Olney, Bucks, MK46 5EJ. Tel: 01234 712023. Fax: 01234 240703. *(Incubators)*

NORFOLK GAME SUPPLIES, Hillside, Wroxham Road, Coltishall, Norwich, NR12 7EA. Tel: 01603 738292. Fax: 01603 738909. *(Incubators)*

SOUTHERN AVIARIES, Tinkers Lane, Hadlow Down, Uckfield, East Sussex, TN22 4EU. Tel: 01825 830283. Fax: 01825 830241. *(Aviaries, meshes, cages, incubators, brooders and other equipment)*

Organisations

WORLD PHEASANT ASSOCIATION, (UK address) PO Box 5, Lower Basildon, Reading, RG8 9PF. Tel: 01734 845140. Fax: 01734 843369. (USA address) 1800 S. Canyon Park. Cir. Ste. 402, Edmond, OK. 73013-6634.

USA Suppliers

MURRAY MCMURRAY HATCHERY, C 56, Webster City, Iowa 50595-0458. *(Range of quails, equipment and supplies)*

ACADIANA AVIARIES, 2500 Chatsworth Road, Franklin, LA 70538. *(Chinese Painted quail)*

STROMBERG'S CHICKS & GAMEBIRDS UNLIMITED, PO Box 400, Pine River, MN 56474.

SUNRISE AVIARIES, 50 Oakwood Road, Orinda, CA 94563. *(Chinese painted quail)*.

QUAIL RUN GAME BIRD RANCH, 5555 Comstock Road, Beaumont, TX 77708.

SEVEN OAKS GAME FARM, 7823 Masonboro Sound Road, Wilmington, NC 28409-2672.

SPRINGHETTI GAMEBIRD FARM, 12220 Springhetti Road, Snohomish, WA 98290. *(Blue Scale, Gambel, Valley)*

Index

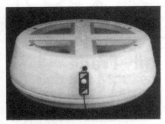